Macmillan Computer Science Series
Consulting Editor
Professor F. H. Sumner, University of Manchester

S. T. Allworth and R. N. Zobel, *Introduction to Real-time Software Design, second edition*
Ian O. Angell, *A Practical Introduction to Computer Graphics*
Ian O. Angell and Gareth Griffith, *High-resolution Graphics Using FORTRAN 77*
R. E. Berry and B. A. E. Meekings, *A Book on C*
G. M. Birtwistle, *Discrete Event Modelling on Simula*
T. B. Boffey, *Graph Theory in Operations Research*
Richard Bornat, *Understanding and Writing Compilers*
J. K. Buckle, *Software Configuration Management*
W. D. Burnham and A. R. Hall, *Prolog Programming and Applications*
J. C. Cluley, *Interfacing to Microprocessors*
Robert Cole, *Computer Communications, second edition*
Derek Coleman, *A Structured Programming Approach to Data*
Andrew J. T. Colin, *Fundamentals of Computer Science*
Andrew J. T. Colin, *Programming and Problem-solving in Algol 68*
S. M. Deen, *Fundamentals of Data Base Systems*
S. M. Deen, *Principles and Practice of Database Systems*
Tim Denvir, *Introduction to Discrete Mathematics for Software Engineering*
P. M. Dew and K. R. James, *Introduction to Numerical Computation in Pascal*
M. R. M. Dunsmuir and G. J. Davies, *Programming the UNIX System*
K. C. E. Gee, *Introduction to Local Area Computer Networks*
J. B. Gosling, *Design of Arithmetic Units for Digital Computers*
Roger Hutty, *Fortran for Students*
Roger Hutty, *Z80 Assembly Language Programming for Students*
Roland N. Ibbett, *The Architecture of High Performance Computers*
Patrick Jaulent, The 68000 – *Hardware and Software*
J. M. King and J. P. Pardoe, *Program Design Using JSP – A Practical Introduction*
H. Kopetz, *Software Reliability*
E. V. Krishnamurthy, *Introductory Theory of Computer Science*
V. P. Lane, *Security of Computer Based Information Systems*
Graham Lee, *From Hardware to Software – an introduction to computers*
A. M. Lister, *Fundamentals of Operating Systems, third edition*
G. P. McKeown and V. J. Rayward-Smith, *Mathematics for Computing*
Brian Meek, *Fortran, PL/1 and the Algols*
Barry Morrell and Peter Whittle, *CP/M 80 Programmer's Guide*

(*continued overleaf*)

Derrick Morris, *System Programming Based on the PDP11*
Pim Oets, *MS-DOS and PC-DOS — A Practical Guide*
Christian Queinnec, *LISP*
W. P. Salman, O. Tisserand and B. Toulout, *FORTH*
L. E. Scales, *Introduction to Non-linear Optimization*
Peter S. Sell, *Expert Systems — A Practical Introduction*
Colin J. Theaker and Graham R. Brookes, *A Practical Course on Operating Systems*
J-M. Trio, *8086–8088 Architecture and Programming*
M. J. Usher, *Information Theory for Information Technologists*
B. S. Walker, *Understanding Microprocessors*
Peter J. L. Wallis, *Portable Programming*
Colin Walls, *Programming Dedicated Microprocessors*
I. R. Wilson and A. M. Addyman, *A Practical Introduction to Pascal — with BS6192, second edition*

Spatial Structure and the Microcomputer

Selected Mathematical Techniques

A. N. Barrett

Lecturer, Department of Computer Science
Brunel University
Uxbridge, Middlesex

A. L. Mackay

Reader in Crystallography
Birkbeck College
University of London

Cheeney
1987 June

MACMILLAN

First published 1987

Published by
MACMILLAN EDUCATION LTD
Houndmills, Basingstoke, Hampshire RG21 2XS
and London
Companies and representatives
throughout the world

Typeset by TecSet Ltd,
Wallington, Surrey

Printed in Great Britain by
Camelot Press Ltd
Southampton

British Library Cataloguing in Publication Data
Barrett, A. N.
    Spatial structure and the microcomputer:
    selected mathematical techniques. —
    (Macmillan computer science series)
    1. Geometry—Data processing
    2. Microcomputers
    I. Title   II. Mackay, A. L.
    516'.0028'5416   QA447

ISBN 0-333-39284-1

Contents

Table of computer programs

Preface

This book is intended for readers with interests in the representation and analysis of spatial structure using the microcomputer.

In almost every branch of science and engineering problems frequently arise which require solutions involving some aspect of spatial structure. For example, the graphical representation of complex three-dimensional surfaces and the improvement of two-dimensional pictorial data for human interpretation are just two problems requiring a combination of structural analysis and computer methods for successful solution. The following chapters present a variety of methods relevant to these and other topics which frequently arise in the course of solving problems involving spatial structure.

The earlier chapters present what is commonly known as the 'real' space approach to spatial structure and involve the use of Cartesian systems to represent structures. The subsequent chapters are concerned with the representation of spatial data in 'transform' space and describe methods frequently used in solving problems which arise in pattern recognition and signal processing.

The level of mathematics presented should be within the grasp of most undergraduate students working in the fields of computing, engineering and the physical sciences. For those readers with less interest in the mathematics than in the practical application of the methods, the book should still appeal since many programs written in BASIC together with documentation and examples illustrating the running of the programs are provided in an appendix and can be used to apply the methods to different sets of data.

It is true to say that microcomputers have become a dominating force in the computing field. Calculations which several years ago would have taxed the resources of many of the larger computers can now be carried out by many of the commercially available microcomputers. Furthermore, with the development of the 16-bit and 32-bit microprocessors, the equivalent computing power is now available at a fraction of the original cost. Clearly, however, not all the problems encountered in problems of spatial structure can be solved using a microcomputer since in extreme cases the extent of the calculations together with the sheer volume of data involved prove unsolvable by even the largest computer yet invented. Nevertheless, many of the principles which involve the methods presented here can be usefully explored and in many cases used to provide solutions to problems of spatial structure.

1 Introduction

Nobody who has experienced the access of personal intellectual power provided by the computer will willingly give it up and go back to the old days of pen, pencil, logarithm tables, setsquare, protractor and eraser for writing, calculating, drawing, planning and designing. Already a number of universities require each student to have his own personal computer and from now on any graduate or professional without his own computing facilities will be at a great disadvantage. The computer is an all-purpose model and a general thinking machine, more because it forces and focuses thought than because it thinks for itself. If you can explain your problem to the computer then you really understand it. The discipline of concise expression in Latin is a soft option in comparison with that of programming a computation.

In the Middle Ages the new symbols for calculation, the Arabic/Hindu numbers, arrived in Europe. They were resisted by the professionals, who had a vested interest in the use of the cumbrous notation of Roman numerals, but spread rapidly among merchants whose livelihoods depended on calculation. The book which carried the information was *Algorithmi de numero Indorum* [(The book of) Al-Khwarizmi on the numbers of the Indians]. Chaucer referred to 'augrims' as the new methods of calculation. Al-Khwarizmi described the process for solving a quadratic equation which has been beaten into every grammar school boy. Ever since then the way or method of carrying out a calculation has been called an 'algorithm' after this author who said: "With my two algorithms, one can solve all problems — without error, if God will!" (*Algebra. Compact Introduction to Calculation by Rules of Completion and Reduction*, Muhammed ibn Musa Al-Khwarizmi (780–850)). The modern version of the algorithm in question is that the solution to $ax^2 + bx + c = 0$ is $x = (-b \pm (b^2 - 4ac)^{\frac{1}{2}})/(2a)$. The new methods of calculation introduced nearly as much revolution in commerce as has the computer in our own times.

Mastery of space is necessary for the application of the computer to the problems of real life. Our own brains have evolved to handle three-dimensional space with inputs derived from our eyes, hands and other senses — those of our ancestors who could not visualise spatial structure and dynamics in three-dimensions fell out of the trees and did not reproduce. The Greeks were good at geometry but did not have the necessary models, which are very tedious to build, for developing three-dimensional geometry. Drawing in the sand is not good enough. Now we have graphics for visual output and devices for visual input to the computer so that our naturally developed senses and abilities can be well matched and complemented by those of the computer. The Classical Tradition, in which our elite has hitherto been educated, has been recently shown to be lop-sided in its neglect of science and technology. In his study *Gears from the Greeks* the late Derek deSolla Price showed that the Anti-Kythera Machine, a block of corroded bronze recovered from a Greek vessel which sank near the island of Anti-Kythera in the first century before Christ, was, in fact, a very complicated calculator for computing the positions of the heavenly bodies (deSolla Price, 1974). It contained some 31 gearwheels, in trains involving epicyclic and differential arrangements and altogether was of the complexity of the mechanical pocket watches used until our own period. We have totally underestimated the technological and practical scientific capabilities of the Greeks because this mechanic tradition has been suppressed by ancient and modern scholastics. Undoubtedly Archimedes himself was familiar with the predecessors of such calculators, but even he was forced to conform to the tradition of affecting to despise the mechanic arts. He said that "certain things first became clear to me by a mechanical method, although they had to be demonstrated by geometry afterwards because their investigation by the said method did not furnish an actual proof." (Archimedes, *The Method*). Computer science is still in the same position and mathematicians are reluctant actually to use computer methods, confining themselves to the ruler and compass methods which were rules of the game fossilised by Euclid.

We regard the availability of the computer as a great liberating force, stirring geometry out of its strait-jacket by introducing new methods and new problems. But we do not any longer, as in the earliest days of the computer, have to try to digest pages of numbers. Instead, we have: Geometry → kinematics → dynamics and Image processing → perception → vision − all in 3-D and in colour rather than just as numbers on paper.

1.1 The Computer

The objective of this book is to see how the geometry of the school books can be carried over to become the basis of a new and exciting field of the computer manipulation of space, the foundation of robotics and of engineering. Euclid's *Elements* [of geometry] has been the second most popular book in the history of the world. It is still of relevance and importance and we wish to show how it is now the foundation of a host of new activities. However, there have been many new developments in geometry taking one far beyond Euclid. In order to become a journeyman in these new technologies it is necessary, as for the would-be carpenter or wheelwright in former days, first to make one's own basic tools. We want to demonstrate how to do this. Only then can one appreciate bought software. We wish to write programs for use as amateurs, and not to deal with all possible contingencies which might cause failure. We are not operating rockets or writing professional software, but writing for ourselves.

When we compute, we use a machine to carry out calculations which are too difficult or too long for us to carry out unaided. It is necessary at all times to keep a sense of proportion and to consider the cost-effectiveness of what we are doing. A common reaction in the early days of learning to compute is "I've done it by computer and it only took three times as long as if I had done it by hand!"

The stages may be something like:

(1) Think out what is to be done and what the value of the job is ("A job which is not worth doing is not worth doing properly" — J. D. Watson).
(2) Think out one or more ways of doing the job — select an algorithm.
(3) Sketch out the program, transposing the problem into computer terms. Assess what storage may be required and estimate the limits of the problem. What are we limited by — time or storage or complexity? What is the rate-determining factor? What are the expected answers? How will we check the results? Do we have particular cases for which we know the answer? How is the problem to be segmented? Can we do it stage by stage?

The importance of hierarchisation cannot be overestimated. H. A. Simon in his book *The Science of the Artificial* (Simon, 1969) gives the examples of two watchmakers who had to assemble watches, each of some 1000 components. Every time they were interrupted, the work in hand fell to pieces and had to be started again from the beginning. The first watchmaker simply inserted the pieces serially and almost never completed the assembly. The second watchmaker, however, assembled sub-units each of ten pieces, then took ten of each of these to make bigger sub-assemblies, and finally put ten of the bigger sub-assemblies to make a watch. This strategy proves incomparably superior in its resistance to disruption. Just the same applies to assembling computer programs which work.

(4) Write the program. As you go, keep appropriate backup copies so that you will not suddenly lose an hour's work or more through carelessly pressing the wrong key or through interference on the electricity supply or dirt on the disc or some such fault.
(5) Make the program work. This is the difficult stage involving systematic fault-finding. To help in this we should arrange that something happens by way of output from the program every minute or so, so that we can check that things are happening correctly. It is better to put in many statements for interim printout and to reduce these later as things prove correct. From intermediate printout we can localise the faults.
(6) Check the program. Does it give the correct answer for simple cases?
(7) Make the definitive runs.
(8) Decide what to keep on disc or on paper and make sure that what you keep is documented so that a stranger coming to it after a long period could make sense of it. The stranger will usually be yourself!

At each of these stages we may have to go back to the beginning and to review the course of the calculation. The whole art of programming is how to keep some sense of proportion so that immense effort is not expended on useless tasks. It is quite easy to set oneself a nearly impossible task. It is clear that the definitive runs, even if they take a long time, like 24 hours, may well be the smallest item in this list, so that this affects our choice of algorithm. It follows that, unless we wish to produce a program which is to be run hundreds of times, a simpler but slower algorithm is strongly to be preferred. If the objective is experience or self-instruction then make sure that we recognise it.

1.2 The Equipment

Computing machinery is changing rapidly so that we do not wish to gear our account to specific machines. We are concerned with the fundamental algorithms which can be implemented on any machine. Already we see a tendency for people to use ready-made computer programs, software, rather than to write their own. We wish to help people to develop their own programs, because only in that way can they achieve the necessary mastery and appreciate the advantages and limitations of bought programs. In computing, as in most other fields, one learns primarily by doing things oneself.

The minimum equipment for a personal system consists of a microcomputer, a display screen, a printer and a disc drive (or other device for holding information in quantity). There are now endless permutations of these components as perusal of the journals devoted to them will reveal. In our view there is no way of learning to swim but going into the water and starting. Since there are now huge numbers of microcomputers about, the best way to begin is to use someone else's installation. The printer which we will select will be determined by the standard of printing which we aim at for letters or text.

1.3 Software

Almost all microcomputers are furnished with some version of BASIC, which was the computer language developed hy John Kemeny and Thomas Kurtz at Dartmouth College, New Hampshire, USA. It was designed so that all students, not just scientists, should be able to compute with satisfaction.

As normally supplied, BASIC is an interpretative language, that is, one statement at a time is translated from the text as written into the internal code of the machine and is executed. The system then looks for the next statement and translates and executes that. It is possible also to take the whole program and translate it all into machine language and to execute the machine code version. This is called compilation. It is possible to obtain a compiler for BASIC for some micros and, if the actual time of calculation is a serious constraint, this may be well worth the money. In particular, in writing graphics programs, where something is drawn on the screen, the slowness of interpreted BASIC may be painfully evident.

We do not here wish to get into the problem of writing programs directly in machine code. Obviously this is specific to each machine. Nevertheless we should be aware that if we are computing seriously, segments of the program may be written in machine code to realise the greatest speed from the machine.

1.4 Functions, Procedures and Sub-routines

We have seen, from the case of H. A. Simon's watchmakers, the great importance of segmenting a program into parts which can be independently built for later assembly as pre-fabricated units into an intelligible structure. Functions, procedures and sub-routines are such independent units. Versions of BASIC differ greatly in what is offered in respect of higher program structure. It is essential to read the manuals of the particular computer, with which we are working, to find out what facilities we have to work with. We may summarise some examples.

Languages differ in their variable types. Microsoft BASIC has double precision, integer and string variables as well as real number variables and these can be allocated by declaration before use. Other languages have variables identified in type by suffix.

Microsoft BASIC

CHAIN to call a program and pass variables to it from the current program using the statement COMMON in the current program. Simple variables, arrays and if necessary all variables can be passed over. It is also possible to have modified COMMON facilities when the programs are compiled with the Microsoft BASIC Compiler.

Function definitions must be confined to a single line (of at most 255 characters). There are no separate procedures or sub-routines (except the GOSUB and RETURN statements). Using variables, like A4, B4, C4 in sub-routine number 4, will help to prevent clashes and will avoid using the same variable twice.

There are IF statements of the form: IF . . . THEN[. . .ELSE] and IF . . . GOTO which may be nested.

BBC BASIC

BBC BASIC has procedures and functions which are defined separately in a section of the program *which is not executed*, that is, which comes after an END or STOP statement. Variables can be passed to a procedure explicitly as

DEF PROCNAME(A,B,C) and other variables will be common to the procedure. A procedure or function may have its own separate variables defined as LOCAL. "LOCAL saves the values of the external variables names and restores these original values when the function or procedure is completed." Arrays cannot be passed to a procedure. Procedures and functions may comprise several lines and may call each other. A procedure is called by PROCNAME(. . .) and a function as X=FNNAME(. . .). A function is defined by DEF FNNAME(. . .) and must contain a line beginning with an equals sign.

Variables A% to Z% are not zeroed and will be passed to a succeeding program by the command CHAIN.

BBC BASIC also has IF. . .THEN. . .ELSE. . ., ON GOSUB, and ON GOTO facilities.

Sinclair QL SuperBASIC

The Sinclair QL version of BASIC has procedures, with LOCAL variables. Procedure calls may be recursive and arrays may be passed to a procedure. Functions go similarly. DEF PROC NAM(. . .) END DEF segments may come anywhere in the program and may be executed. There are also multi-line IF statements: IF. . .THEN. . .ELSE. . .END IF, and these may be nested, although confusion may arise as to which ELSE corresponds to which IF. ON. . .GOTO and ON. . .GOSUB statements are provided.

SELECT ON. . . END SEL is still another program structure which is provided.

1.5 Errors, Mistakes, Blunders and Inaccuracies

The whole question of error is of dominating importance in computing. One has only to read of the failure of the launch of a space rocket because of a comma in the wrong place, to realise that the whole future of mankind on this planet may turn on a computer error. Fortunately most of our mistakes do not have such grave consequences and we can expend less effort on making programs which check everything. In most cases the success or otherwise of a program is self-evident.

It is this test which distinguishes the scientific and technical culture from the literary. If you build a bridge or a computer program success or failure is clear. If you build a system of philosophy or a novel then there is no such test. We may simply list some relevant considerations:

(1) If you type a text, even working carefully so that you make only one mistake every four lines, then you have only a one in a thousand chance of completing a page (of 25 lines) without a single mistake. Thus, error cannot be eliminated and our strategy must be to find and correct errors

as effectively as possible. The first line of action is simple. Learn to type efficiently. Most computer people are hopeless in this respect and use only two fingers. The greatest gain in speed in computing in general can be achieved in this department.

(2) Since the greatest source of mistakes is oneself, then the principal remedy must be self-discipline. It is necessary to develop constant habits on various scales, from the writing of individual statements to the design of whole suites of programs, so that you can read your own program as a grandmaster scans the chessboard, seeing the consequences of each part of the situation. Emanuel Lasker, a grandmaster, speaking about chess, said what applies as much to the computer programmer: "On the chess board, lies and hypocrisy do not survive long. The creative combination lays bare the presumption of a lie; the merciless fact, culminating in a checkmate, contradicts the hypocrite."

(3) We have to recognise and arrange to live with our own failings, such as our memory, so that for significant programs, which you intend to keep, it is necessary to write as if for a stranger, unfamiliar with your mental habits, who may take up the program a year or two hence. Put in enough comments to enable a stranger to reconstruct the main lines of operation.

(4) From time to time it is essential to read and to re-read the manuals of the hardware and software which you use. On a first or second reading it is impossible to take in everything and many manuals are notoriously badly written. Manuals rarely say anything which is untrue, but the truth does not always leap out. There are occasional bugs (mistakes) in commercial software widely available.

(5) Get accustomed to writing modules which do specific unit tasks and then slot them in, in pre-fabricated form, to make bigger programs. This, of course, is what sub-routines are supposed to do and why sub-routines which can be developed quite independently with local variables are so important. It is this technique alone which makes complex programs possible.

Inaccuracies

There are many books on the estimation of probable errors in numerical calculations which is often a very technical matter.

In general we do not have to worry very much about the accuracy of calculation, especially if the results are shown graphically. However, we have to keep in mind that, for example, the quotient $(a - b)/(c - d)$ becomes progressively indeterminate as the two differences get smaller. This is a singularity and our programs may encounter such traps in various forms. We may cover the case of $c > d$ and the case of $c < d$ but we must also think of what will happen if $c = d$. This takes us into the question of the representation of integers and of real numbers in our particular system. Many mathematical functions have singularities, where their values are infinite or undetermined. For example, the tangent of $90°$ is infinite so that a program involving it may crash if the argument happens to go close to $90°$.

In computer graphics trouble may arise in rotating an object by successive multiplication of the x, y, z coordinates with a rotation matrix. If we concatenate the operations of rotation then some of the coordinates will from time to time be very small but will grow again as we turn further. Thus, since we cannot regain significant digits once lost, the general accuracy will be progressively eroded. The use of homogeneous coordinates $(x, y, z, 1)$ and 4×4 rotation matrices is designed to avoid this difficulty.

In a number of areas test data is available. For example, there are books of standard test examples for the inversion of matrices and for the calculation of eigenvalues which are important problems where programs may have to deal with a variety of 'pathological' cases.

Blunders

This is really the gravest danger which the amateur computer operator faces. It is extremely easy to become confused and to erase a huge volume of work by mistake. Only making a few such blunders will provide the incentive to keep backup copies and records of what is happening in a systematic way.

Misconceptions

The most difficult kind of error to find is one where we have got the whole method wrong. The only cure is to keep checking wherever possible against known results.

2 Mathematical processes for computation

We look first at the process of transferring from algebra to a computer language. We must begin by examining our chosen programming language to see what functions are supplied. Some languages such as Pascal are poor in supplied functions, and everything must be written afresh, but usually we have a good range, both of algebraic functions, like trigonometric functions — sines, tangents, cosines and their inverses — and also of manipulative functions, like maxima and minima, integer parts, etc. Note that there are minor dialectal differences, such as ATN or ATAN, SQR or SQRT, etc. which can cause delays in transcription.

Special functions may have to be written as routines. Many functions can be written as polynomials (tables are available) and indeed functions such as sines are evaluated in machine language by Chebyshev polynomials which are polynomials with coefficients carefully chosen to give the necessary accuracy (and no more) with the minimum number of terms. Korn and Korn (1968), for example, give tables.

Inverse sines and cosines may have to be derived from the inverse tangent and defined as user functions. For example, the inverse sine may be written as

DEF FNS(X)=ATN(X/SQR(1−X∗X))or DEF FNS(X)=2∗ATN((1−SQR(1−X∗X))/X) (the first function will fail at X=+1 or −1 and the second if X=0

and the inverse cosine as

DEF FNC(X)=ATN(SQR(1−X∗X)/X) + (1−SGN(C))∗PI/2 or
DEF FNC(X)=2∗ATN(SQR((1−X)/(1+X))) (where the first will fail for X=0 and the second for X=−1)

If a function is a polynomial $Y = A + BX + CX^2 + DX^3$ ... we may compute it as $Y = A + X*(B + X*(C + X*D))$ ... which avoids taking exponents, but $Y = A + B*X + C*X*X + D*X*X*X$ is nearly as good. There are many devices for going faster but usually clarity is better than speed or economy of storage. We may gradually adopt a consistent style which will help greatly in making our mistakes visible.

2.1 Integration

In order to find the area under a curve, such as one given by $y = f(x)$, the area has to be divided into elementary rectangles and the areas of these are summed. When such calculations were done by hand, many devices were used to keep down the time needed for the calculations but now, when programming time is usually much longer than the actual computer time, we can afford to be more prodigal with the latter and the simplest rules will usually suffice.

For example, in taking the areas of elements it is enough to take the height y at the midpoint of the slice and to neglect curvature of the line over the element

$$\int_{x=x_1}^{x=x_2} y \, dx = \sum_{i=1}^{i=N} y_i \, dx$$

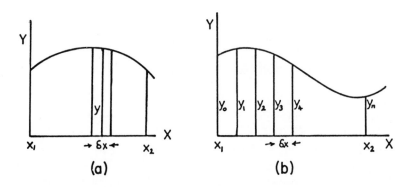

Figure 2.1. (a) Integration is the determination of the area under a curve of $y = f(x)$ plotted against x for $x = x_1$ to $x = x_2$. The area is the sum of the areas of strips of width δx and height y (measured at the midpoint of the strip). (b) Simpson's Rule, area $= \delta x(y_0 + 4y_1 + 2y_2 + 4y_3 + \ldots + 2y_{n-2} + 4y_{n-1} + y_n)/3$ fits a parabola to the curve in each element.

Simpson's Rule corresponds to fitting a parabola to the profile of each element

$$\int_{x=x_1}^{x=x_2} y \, dx = \delta x \, (y_0 + 4y_1 + 2y_2 + 4y_3 + \ldots + 2y_{n-2} + 4y_{n-1} + y_n)/3$$

[n should be even]

More complex rules are to be found listed in the manuals. We may take the calculation of the error function as an example:

$$\text{erf}(x) = (2/\sqrt{\pi}) \int_{t=0}^{t=x} \exp(-t^2) \, dt$$

```
10 REM calculate error function
20 DEFINT I
```

```
30 DEFINT N
40 PRINT "input argument"
50 INPUT A
60 PI=3.1415926535
70 F=0
75 REM number of intervals
80 N=500
85 REM interval of integration
90 E=A/2/N
100 FOR I=0 to N−1
110 X=E+A*I/N
120 D=EXP(−X*X)
130 F=F+D
140 NEXT I
150 F=F*A/N*2/SQR(PI)
160 PRINT "error function"
170 PRINT "x,    erf(x),    last increment"
180 PRINT A,F,D*A/N*2/SQR(PI)
```

If we had wished to be more sophisticated we could have arranged to stop iteration when we had reached a sufficient accuracy.

2.2 Derivatives

If F is an algebraic function, for example a polynomial, such as $F = AX^3 + BX^2 + CX + D$, then its first derivative, its slope, is found algebraically using the standard rule $(d/dx)(x^n) = nx^{n-1}$, so here $dF/dX = 3*A*X*X + 2*B*X + C$. This corresponds to taking the limit of $[F(x + \delta x) - F(x)]/\delta x$. However, if the function F is difficult to differentiate algebraically, we can always perform the calculation numerically, taking δx as a suitably small interval. Again, many manuals of integrals and derivatives are available.

2.3 Complex Numbers

Complex numbers are an algebraic device for carrying out geometrical operations in two dimensions. This type of algebra was invented by Caspar Wessel about 1797. By using the symbol i, the only property of which is that $i^2 = -1$, pairs of numbers can be manipulated as single entities. Plane geometry can be carried through using complex numbers (for example, see Zwicker (1950)). This may appear to be no great advance on ordinary ways of handling (x, y) coordinate pairs, but there are other applications. Some computer languages have special complex variable types. We will not use this algebra for geometry but as a precursor for the more complicated algebra of the vectors and quaternions which are essential for the three dimensions of space. However, a wave is described by an amplitude and a phase angle with respect to a starting point, and a wave falling on a screen thus requires two quantities to be specified at each point. This pair may be either amplitude and phase or cosine and sine components. Thus, complex numbers are essential for image processing, where pictures are made up by superimposing waves of density, which must be added in the correct phase. They are also necessary for alternating current networks where 'in phase' and 'in quadrature' components have to be handled separately — these are the same cosine and sine components. A sine wave of arbitrary phase with respect to a given starting point can be expressed as the sum of one sine wave and one cosine wave which start with zero phase from the given origin. The sine wave is anti-symmetrical with respect to the origin and the cosine wave is symmetrical. (That is, changing the sign of the argument of a sine wave inverts it, whereas changing the sign of the argument of a cosine wave leaves it unchanged.)

The basic feature is that every equation in complex numbers is really two equations. Each term can be separated into 'real' and 'imaginary' parts and these can be separately equated — the process being equivalent to the resolution of forces in two perpendicular directions. Two mutually perpendicular axes are used. Ordinary units are used for measuring along the X-axis and units of i (or, among engineers, j) are used for coordinates along the Y-axis. The symbol z is often conventionally used for a complex number so that one can write $z = x + iy$. z^*, the complex conjugate of z, is given by $z^* = x - iy$ and is a quantity which is frequently required. We can always express n linear equations in complex variables as $2n$ ordinary linear equations. Non-linear equations in complex quantities are more difficult to separate into real and imaginary parts.

If $z = x + iy$, then the real and imaginary parts are $R(z) = x$ and $I(z) = y$. Complex numbers can be added, subtracted, multiplied, divided and raised to a power like ordinary quantities, but if we wish to calculate the results numerically the answers must be separated into real and imaginary parts. If we have

$$z = z_1 + z_2$$

then

$x = x_1 + x_2$ and $y = y_1 + y_2$ and, for example,

$$z_1 z_2 = (x_1 x_2 - y_1 y_2) + i(x_2 y_1 + x_1 y_2)$$

If z is considered as the vector (x, y) then its amplitude or length is $(x^2 + y^2)^{1/2}$. The amplitude r of any complex quantity is obtained by multiplying by its complex conjugate and taking the square root of the product, which is necessarily real. Thus $r^2 = x^2 + y^2$. The phase angle is θ so that $z = x + iy = r\cos\theta + ir\sin\theta$ and $\tan\theta = y/x$. This enables us to use the relation $\exp(i\theta) = \cos\theta + i\sin\theta$.

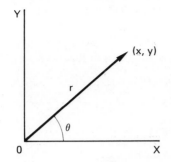

Figure 2.2. Polar coordinates (r, θ) are related to Cartesian coordinates (x, y) by $x = r\cos\theta$, $y = r\sin\theta$, $r = (x^2 + y^2)^{1/2}$, $\theta = \arctan(y/x)$.

Thus, rotation of the vector z through an angle θ is accomplished by multiplying z by $\exp(i\theta)$. This multiplication changes the phase of z but not its amplitude. The complex number algebra gives us a means of adding waves (of the same frequency in time or space) which may have arbitrary phase. This is essential for alternating current theory and for image processing. The operations of matrix manipulation can be written for complex variables and this is necessary for electrical network calculations with alternating currents.

Versions of BASIC which have independent sub-routines with local variables (such as, for example, the versions of PROCEDURES implemented on the Sinclair QL), permit one conveniently to write sub-routines to perform the standard operations of scalar arithmetic with complex numbers. At little further complication these can be carried over to the manipulation of arrays of complex numbers. We have, for example, to find $z_1 z_2$, z_1/z_2, $1/z$, $z\hat{\ }n$, $z_1\hat{\ }z_2$ — in each case we define functions for the real and the imaginary parts. (We may use the sign $\hat{\ }$ as 'raised to the power of', as is conventional on computer keyboards.)

For the product of two complex numbers (A+iB) and (C+iD)

```
DEF FNPR(A,B,C,D)=A*C−B*D:   REM real part of product
DEF FNPI(A,B,C,D)=B×C+A*D:   REM imaginary part of product
```

For the reciprocal of (A+iB)

```
DEF FNRR(A,B)=A/(A*A+B*B)
DEF FNRI(A,B)=−B/(A*A+B*B)
```

For the value of (A+iB)^N

```
DEF FNNR(A,B,N)=(A*A+B*B)^(N/2)*COS(N*ATN(B/A))
DEF FNNI(A,B,N)=(A*A+B*B)^(N/2)*SIN(N*ATN(B/A))
```

Here, of course, we must avoid the case of A=0, testing for it separately, where, if N is an even integer the value will be wholly real, and if N is an odd integer, wholly imaginary. The real part will then be (B^N)*COS(N*PI/2) and the imaginary part (B^N)*SIN(N*PI/2).

For the value of (A+iB)^(C+iD), the real part of (A+iB)^C is FNNR(A,B,C) and the imaginary part is FNNI(A,B,C) so that we need only consider (A+iB)^(iD).

DEF FCER(A,B,D)=cos((A*A+B*B)^(D/2))*EXP(−D*ATN(B/A))

and

DEF FCEI(A,B,D)=sin((A*A+B*B)^(D/2))*EXP(−D*ATN(B/A))

Thus (A+B)^(C+iD) has a real part

FNNR(A,B,C)*FCER(A,B,D) − FNNI(A,B,C)*FCEI(A,B,D)

and an imaginary part

FNNR(A,B,C)*FCEI(A,B,D) + FNNI(A,B,C)*FCER(A,B,D)

For the square root of (A+iB), the real part is

DEF FSQI(A,B)=SQRT(−0.5*A+0.5*SQRT(A*A+B*B))

and the imaginary part

DEF FSQI(A,B)=SQRT(−0.5*A+0.%*SQRT(A*A+B*B))

2.4 Vectors

The algebra next more complicated than that of *complex numbers* is that of *vectors*, where we have three quantities, i, j and k, rather than two, which are themselves unit vectors respectively in the directions of the three Cartesian axes, OX, OY, OZ. It is conventional to print a *bold* symbol for a vector, as *r* (in handwritten texts, or where a bold typeface is not available, it is usual to underline the vector symbol, as r̲). Just as a complex number equation implies that real and complex quantities are separately equal, that is, comprises two equations, so a vector equation means that components resolved in three non-coplanar directions are equal, that is, it gives three equations. A unit vector has projections of lengths l, m, n on the three orthogonal axes OX, OY and OZ. These are called the direction cosines of the vector and $l^2 + m^2 + n^2 = 1$.

A vector is a quantity with magnitude and direction in space but no position (that is, it may be displaced arbitrarily as regards point of action). Not all quantities which can be represented as directed magnitudes are vectors. The criterion is that they should combine as vectors. Finite rotations can be represented as directed magnitudes but they do not add as vectors. In the computer vectors are handled as one-dimensional arrays. Many texts (such as Hague (1970) and Weatherburn (1924, 1935, 1939)) describe vector algebra, so that we need here include only a sketch.

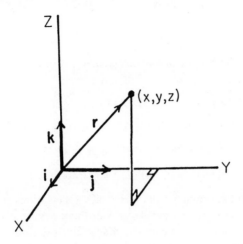

Figure 2.3. A vector V from the origin to the point (x, y, z) can be expressed as the sum of components along OX, OY, OZ in terms of the unit vectors i, j, k so that $V = V_x i + V_y j + V_z k$.

$$r = r_x i + r_y j + r_z k$$

where i, j and k are unit vectors along the orthogonal coordinate axes OX, OY and OZ.

$r = s$ implies

$$r_x = s_x, r_y = s_y, r_z = s_z$$

The amplitude or length of r is

$$|r| = (r_x^2 + r_y^2 + r_z^2)^{1/2}$$

$(r + s)$ has components

$$r_x + s_x, r_y + s_y, r_z + s_z$$

along i, j and k.

There are two products, the scalar or inner product, written $a.b$ which is a number $(a_x b_x + a_y b_y + a_z b_z$, also equal to $|a| |b| \cos(\gamma_{ab})$ where γ_{ab} is the angle between the vectors a and b), and the vector product, written $a \times b$.

We give routines in BASIC for *inner product, scalar product, vector product, outer product* and *triple product*.

For two vectors $A<3>$ and $B<3>$ the inner product or scalar product is the scalar number S, such that, S=A(1)*B(1)+A(2)*B(2)+C(3)*C(3). The vector product is a vector $C<3>$ such that

$$C(1) = A(2)*B(3) - A(3)*B(2)$$
$$C(2) = A(3)*B(1) - A(1)*B(3)$$
$$C(3) = A(1)*B(2) - A(2)*B(1)$$

The triple product, the volume of a parallelepiped defined by the vectors A<3>, B<3> and D<3> is

$$V = A(1)*B(2)*D(3) + A(2)*B(3)*D(1) + A(3)*B(1)*D(2)$$

$$- A(1)*B(3)*D(2) - A(2)*B(1)*D(3) - A(3)*B(2)*D(1)$$

The outer product of A<3> and B<3> is an array C<3,3>, where

$$C(i,j) = A(i,1)*B(1,j) + A(i,2)*B(2,j) + A(i,3)*B(3,j)$$

Vectors are useful in geometry and mechanics. Calculations are made first with vector symbols, free of any particular coordinate system, and only at the end are they converted to x, y, z components for the insertion of particular numerical values. Vectors may also be seen to be a special class of more general quantities called tensors.

Vectors can be differentiated and the whole of differential geometry can be concisely expressed.

3 Matrices

The whole point of the computer is that it can carry out mathematical operations at great speed. Some of the most tedious but necessary operations are concerned with the handling of matrices, which are arrays (or tables), sometimes very large, of numbers. It is absolutely necessary to be able to handle arrays, and the central feature of the original version of BASIC, designed by John Kemeny and Thomas Kurtz at Dartmouth College, New Hampshire, USA for the use not only of science students but of all students, was this facility. It is very retrograde that most implementations of BASIC for microcomputers are emasculated without this ability.

For example, the equation [A] [X]=[H], where [A], [X] and [H] are arrays, may represent 20 simultaneous linear equations for 10 unknowns [X]. We could insert array dimensions as A<20,10>*X<10,1>=H<20,1>. These may be solved by 'least squares' (see section 3.5) for the best values of [X], as X = [[A^T] [A]] $^{-1}$ [H] in the original version of BASIC as follows (after the arrays have been defined)

MAT B = TRN(A):   MAT C = B*A: MAT D = INV(C):   MAT X = D*H

Here the dimensions would be B<10,20>, C<10,10>, D<10,10>, H<20,1>, A<20,10>, X<10,1>.

The same program in FORTRAN or in an impoverished BASIC has some ten times the length. The ability to do algebra with arrays as easily as with single numbers means that you can write quite complicated programs directly on to the machine as you think out what you want to do, and it is hardly necessary to use software or other people's programs.

Simultaneous linear equations arise in very many contexts and, repeated very many times by computer, their solution can also be used to solve non-linear and other more complicated problems.

We will develop the various methods step by step, introducing the necessary notation. The simplest simultaneous linear equations, for the unknowns x_1 and x_2 are

$$a_{11}x_1 + a_{12}x_2 = h_1$$

$$a_{21}x_1 + a_{22}x_2 = h_2$$

which we may write [A] [X]=[H], with square brackets to show that the quantities are arrays or as A<2,2>X<2,1>=H<2,1> to show the dimensions of the arrays. We may also, in the actual program, use A(2,2) to denote the element in the second row and the second column of the matrix [A]. However, in this printed text we will use subscripts, as in a_{22}, for this element by itself.

These are the simplest simultaneous linear equations – the unknowns are numbered by subscript $x_1, x_2 \ldots$, the right-hand sides $h_1, h_2, \ldots$ similarly. The coefficients form a matrix

$$[A] = \begin{bmatrix} a_{11} & a_{12} \\ a_{21} & a_{21} \end{bmatrix}$$

(We will use the (almost) universal convention that the first suffix numbers the row and the second the column. If all the terms a_{ij} are equal to the corresponding a_{ji}, the matrix is called symmetrical.)

The solution to the set of equations is normally found by the arithmetic process of the elimination of one variable by multiplying the first equation by a_{22}, the second by a_{12} and subtracting

$$a_{11}a_{22}x_1 + a_{12}a_{22}x_2 = a_{22}h_1$$

$$a_{12}a_{21}x_1 + a_{12}a_{22}x_2 = a_{12}h_2$$

$$(a_{11}a_{22} - a_{12}a_{21})x_1 = a_{22}h_1 - a_{12}h_2$$

thus

$$x_1 = (a_{22}h_1 - a_{12}h_2)/(a_{11}a_{22} - a_{12}a_{21})$$

and similarly

$$x_2 = (-a_{21}h_1 + a_{11}h_2)/(a_{11}a_{22} - a_{12}a_{21})$$

When there are many unknowns, for example if there are 20 equations for 20 unknowns x_1, then, although simple in principle, the solution of the set of equations would take a long time and there would be a very high probability of error. Cramer's rule organises the solution as the quotient of two $N \times N$ determinants and we could use the program DETER (see appendix) for evaluating these.

This example of 2 equations for 2 unknowns shows several features common to all such sets of simultaneous equations. (a) The denominator in the expressions for x is the determinant of the matrix [A], which must be square. Such a quantity occurs in all such sets of equations. If the value of the determinant is zero, then there is no solution. Also, if the matrix [A] is not square, there is no immediate solution. (b) We note also that whether or not there is a solution does not depend on [H], the right-hand sides of the set of equations, but only on [A].

3.1 Determinants

It is thus necessary to be able to calculate expeditiously the determinant of a square matrix of $N \times N$ terms. The determinant, when multiplied out, is a single number. If $N = 2$ the value is, as we have seen, $D = a_{11}a_{22} - a_{12}a_{21}$ and if $N = 3$ then the value is

$$D = a_{11}a_{22}a_{33} + a_{12}a_{23}a_{31} + a_{13}a_{21}a_{32} - a_{11}a_{23}a_{32} - a_{12}a_{21}a_{33} -$$

$$a_{13}a_{22}a_{31}$$

This is best remembered by a mnemonic, writing the matrix twice

For these cases it is best, in a program, to write the expression explicitly without the complication of writing loops. In good versions of BASIC the determinant is available as a function. In the general case D, the value of the determinant has $N!$ terms, each of N values multiplied together, so that evaluation by multiplying out is prohibitive. The expression is: $D = E_{ijk}E_{lmn}a_{il}a_{jm}a_{kn}$ for the case $N = 3$, where E_{ijk} (the alternating tensor) has the value 1 if i, j, k are in cyclic order, -1 if they are in anti-cyclic order and 0 if any two are equal, and similarly for higher orders. Summation is made over the repeated indices giving each the values 1, 2, 3 in turn.

A determinant is best evaluated by the process of pivotal condensation, that is, by reducing all terms but one in a row to zero (by subtracting other rows) so that the order of the determinant is reduced by one. This procedure is then carried out recursively. A program (DETER) is given in the appendix. For our present purposes it is not essential to look further into the properties of determinants.

3.2 Rank

A square matrix [A] of $N \times N$ entries (order N) has a rank M, an integer, which may be less than or equal to N. If the determinant of [A] is non-zero, then its rank is N. If the determinant of [A] is zero, then the largest non-zero determinant which can be made from [A] by striking out rows and columns leaves an $M \times M$ array and the rank of [A] is M. The same process may also be applied to a rectangular matrix, which also has a rank. We will see later how practically to find the rank of a matrix using the program GENINV.

When a set of linear equations [A] [X]=[H] has a matrix of coefficients [A] which is rectangular or which has a zero determinant, the ordinary process of solution (by elimination) fails and a program will crash, but all is not lost. For example, if we have two equations $x_1 + x_2 = 1$ and $2x_1 + 2x_2 = 2$ then the determinant is zero and there is no solution. But we at least know that $x_1 + x_2 = 1$ and this information need not be wasted. The rank of the matrix [A] is here not 2 but 1.

If the determinant of [A] in a matrix equation [A] [X]=[H] is zero the matrix is said to be *singular*. A matrix equation, or a set of linear equations which do have a solution, that is, the matrix of coefficients is non-singular, is solved because it is possible, for each non-singular matrix [A] to find an inverse matrix $[A^{-1}]$ such that $[A^{-1}]$ [A]=[I] where [I] is a unit matrix (1s on the diagonal and 0s otherwise). The program INVERT does this. Then, multiplying both sides of [A] [X]=[H] by $[A^{-1}]$ we have [X] = $[A^{-1}]$ [H].

3.3 Multiplication of Matrices

Here we need to be clear how to multiply matrices. Pre-multiplication and post-multiplication are not the same, so that the order is vital. The first requirement is that the matrices should be of conformable dimensions. It is very convenient in sketching out a program to write for each matrix [A] its dimension, A<rows, columns>. [C]=[A] [B] is then $C<i,j>=\Sigma_k A<i,k>*B<k,j>$. Einstein introduced a convention into algebra that, if an equation written with suffixes contained suffixes repeated on the same side of the equation, as in $C_{ij} = A_{ik}B_{kj}$, then summation was to be made over all the values of the repeated suffix; for example C(1,3)=A(1,1)*B(1,3) + A(1,2)*B(2,3) + A(1,3)*B(3,3). In BASIC we would then write, to find [C]=[A] [B]

```
L=8: M=10: N=13:
DIM C(L,M) ,A(L,N) ,B(N,M)
FOR I=1 TO L: FOR J=1 TO M
P=0: FOR K=1 TO M
P=P + A(I,K)*B(K,J): NEXT K
C(I,J)=P: NEXT J: NEXT I
```

In the original (Kemeny and Kurtz form of) BASIC, having allotted dimensions, we would simply write MAT C=A*B.

3.4 Inversion

The matrix $[A^{-1}]$, the inverse of [A], can be found in a number of ways:

(1) Explicitly, by finding the adjoint matrix, having calculated the determinant and found it to be non-zero. The element $A^{-1}(i,j)$ in the inverse matrix is found by taking the element $A(j, i)$ (note the reversal of suffixes) and striking out the row j and the column i through it. $A^{-1}(i,j)$ is then equal to the determinant of the remaining $(N - 1) \times (N - 1)$ matrix, divided by D and attaching the sign $(-1)^{i+j}$ to it. For a 2×2 matrix we may write this explicitly as

$$[A^{-1}] = \begin{bmatrix} a_{22}/D & -a_{21}/D \\ -a_{12}/D & a_{11}/D \end{bmatrix}$$

Thus, if we multiply $A^{-1}<2,2>$ by $H<2,1>$ we have

$$X<2,1>=A^{-1}<2,2>*H<2,1>$$

and

$$x_1 = (a_{22}H_1 - a_{21}H_2)/D$$

$$x_2 = (-a_{12}H_1 + a_{11}H_2)/D$$

as was found above.

For a 3×3 matrix the inverse $[B]$ is given by

$$b_{11} = \quad (a_{22}a_{33} - a_{23}a_{32})/D$$

$$b_{12} = -(a_{12}a_{33} - a_{32}a_{13})/D$$

$$b_{13} = -(a_{12}a_{23} - a_{22}a_{13})/D$$

$$b_{21} = -(a_{21}a_{31} - a_{31}a_{23})/D$$

$$b_{22} = \quad (a_{11}a_{33} - a_{31}a_{13})/D$$

$$b_{23} = -(a_{11}a_{23} - a_{21}a_{13})/D$$

$$b_{31} = -(a_{21}a_{32} - a_{31}a_{22})/D$$

$$b_{32} = -(a_{11}a_{32} - a_{31}a_{12})/D$$

$$b_{33} = \quad (a_{11}a_{22} - a_{12}a_{21})/D$$

and it is as easy to write this out in full as to use a general inversion program.
(2) We can use the Gauss–Jordan method of inversion, for which a program INVERT is given in the appendix.
(3) The inverse matrix can be calculated iteratively by starting with a first guess and obtaining an improved value. The suffix k denotes the kth approximation

$$[B]_{k+1} = [B]_k * [2*[I] - [A]*[B]_k]$$

We will use this method intensively later. The point is simply that we should have a program segment, which can be built into a program whenever we want it, to invert a non-singular matrix. The original version of BASIC does this with the single statement MAT B = INV(A). When writing a program for the first time it is useful to check the inversion by finding $[A^{-1}]\,[A]$ which should be the unit matrix $[I]$.

3.5 Least Squares

There are two very important steps beyond finding the solution to a set of linear equations by inverting a non-singular matrix. The first is the problem of finding the 'best' solution to a set of inconsistent linear equations. The equations may be the results of experiments with inaccuracies of measurement. What we may mean by 'best' has to be defined exactly and there are various choices. The usual one is that the sum of the squares of the discrepancies, when calculated values are put into the equations, should be a minimum. This we will call the 'least squares solution'.

$$a_{11}x_1 + a_{12}x_2 + \ldots + a_{1n}x_n - h_1 = Q_1$$

Q_1 is the discrepancy for the first equation and the least squares solution for the x_1 makes the sum of $(Q_1)^2$ to be a minimum.

We start with more equations, the observational equations, than unknowns, and thus must combine them in some way. The equations are weighted, if necessary, by being multiplied right through by some number. They are then multiplied by the transpose of $[A]$, denoted by $[A^T]$, where each element i,j of $[A]$ is exchanged with the element j,i

$$[A^T]\,[A]\,[X] = [A^T]\,[H]$$

It is convenient to keep track of the process by including the dimensions of the arrays. We have N equations for M unknowns and $N > M$. The equations are

$$A{<}N{,}M{>}*X{<}M{,}1{>}{=}H{<}N{,}1{>}$$

Then

$$[A^T{<}M{,}N{>}A{<}N{,}M{>}\,]X{<}M{,}1{>} = A^T{<}M{,}N{>}*H{<}N{,}1{>}$$

$[A^T{<}M{,}N{>}A{<}N{,}M{>}]$ is a square symmetric matrix of order M so that N equations have been reduced to M. $[A^T A]$ is then inverted (by the process described above) so that

$$X{<}M{,}1{>} = [A^T{<}M{,}N{>}*A{<}N{,}M{>}]^{-1}A^T{<}M{,}N{>}*H{<}N{,}1{>}$$

The discrepancies can if necessary be found from the original equations by

$$A{<}N{,}M{>}*X{<}M{,}1{>} - H{<}N{,}1{>} = Q{<}N{,}1{>}$$

3.6 The Generalised Inverse

Suppose, however, that we have fewer equations than unknowns and $N < M$

$$A<N,M>*X<M,1> = H<N,1>$$

If both sides of the equation below are multiplied by $A<N,M>$ we obtain the original equation

$$X<M,1> = A^T<M,N>[A<N,M>A^T<M,N>]^{-1}H<N,1>$$

$[AA^T]$ is square and is assumed to be non-singular (its determinant is not zero).

Here the expression $A^T[A<N,M>A^T<M,N>]^{-1}$ is called the *generalised inverse* of $[A]$ and may be denoted by $[A^+]$. The iterative program GENINV (see appendix) deals also with the case where $[A]$ $[A]$ is singular.

The generalised inverse is an extremely useful device because every matrix $[A]$, square or rectangular, singular or non-singular, has a unique generalised inverse. Thus, every matrix equation $[A][X]=[H]$ has a solution $[X]=[A^+][H]$. If there is a unique exact solution, $[X]=[A^+][H]$ gives it. If there are more equations than unknowns, then it gives the unique least squares solution; if there are fewer equations than unknowns it gives a solution, but not a unique one. In fact this solution is one of a range and the full range of possible solutions is

$$[X] = [A^+] + [[I] - [A^+][A]][Z]$$

where Z is an arbitrary vector (of dimensions $M,1$). Putting in the dimensions of the arrays

$$X<M,1> = A^+<M,N>*H<N,1> + [I(M,M) - A^+<M,N>A<N,M>] Z<M,1>$$

and

$$M>N$$

The program also gives the rank of $[A]$.

This position needs fuller investigation because this procedure will apparently accomplish the impossible and solve two equations for three unknowns. For $A<3,3>*X<3,1>=H<3,1>$, $[A]$ may be an operation of projection, so that a position x, y, z is projected down on to the plane $z = 0$. Then all knowledge of the z-coordinate is lost and the operation cannot be reversed. However, the knowledge of x and y is not lost and if one insists on reversing the operation of projection, x and y can be recovered but no information about z is available so that z remains arbitrary.

A program GENINV is given in the appendix for finding the generalised inverse $[A^+]$. In special circumstances where $[A][A^T]$ is non-singular A^T $A^T[AA^T]^{-1}$ may be used for the generalised inverse where the inversion is done with a non-iterative algorithm, such as that given as INVERT.

A major use of the generalised inverse is in molecular geometry, where a molecule composed of N atoms has $3N$ coordinates, but where there are only $3N - 6$ independent parameters. (The molecule can be translated or rotated with six degrees of freedom without affecting its shape.)

With routines for finding the determinant, the inverse and the generalised inverse of a matrix, as well as for multiplying matrices, we have, except for the 'eigenvalue problem', all the principal tools for numerical matrix algebra.

3.7 Eigenvalues and Vectors

The Eigenvalue Problem (from the German *Eigenwert* = characteristic value) is that of finding solutions for the equation $[A] [X] = \lambda [X]$, where $[A]$ is a matrix (an $N \times N$ two-dimensional array), $[X]$ is a vector (an $N \times 1$ one-dimensional array) and λ is a constant. The values of λ which permit a solution are called the eigenvalues. There are N of them but some may be the same as others. For each eigenvalue there is an eigenvector — a value for the vector $[X]$ which is taken to have unit length (the sum of the squares of the components is one). There are some complications if any two eigenvalues are the same and the eigenvectors are then not unique.

It is easiest to see what happens in terms of an example. We take the calculation of the principal moments of inertia of a number of mass points about the centre of gravity of the system. The moment of inertia is a physical quantity which is, in fact, a second-order symmetrical tensor. This means that six values are necessary for its specification and these values depend in a characteristic way on the axes we choose with respect to which to calculate the moments. For a single point the moment of inertia about an axis of rotation is the product of its mass and the square of the perpendicular distance from the point to the axis.

If a sphere is distorted into an ellipsoid then the three eigenvectors of the transformation are those directions which are unchanged but where the lengths are multiplied by λ_1, λ_2 and λ_3, the three eigenvalues.

If the axes are chosen correctly, the moments of inertia take a particularly simple form and we then have to consider only the principal moments of inertia which are the moments of inertia for rotations about these axes. These principal axes are, in fact, the eigenvectors of the matrix of the inertia tensor. We will see that the moments of inertia of a general body (or an assembly of points) can be represented as an ellipsoid and that finding the eigenvalues and eigenvectors of this ellipsoid corresponds to finding the lengths and directions of the principal axes of this ellipsoid.

The inertia tensor [M] is the matrix

$$\Sigma \begin{bmatrix} (my^2 + mz^2) & -mxy & -mxz \\ -myx & (mx^2 + mz^2) & -myz \\ -mzx & -mzy & (mx^2 + my^2) \end{bmatrix}$$

or

$$[M] = -\Sigma mx_i x_j + (\Sigma mr^2)\,[I]$$

where $r^2 = x^2 + y^2 + z^2$.

If $\Sigma(mr^2) = \Sigma(m)R^2$, then R is called the radius of gyration of the system. If the origin is taken at the centre of gravity of the system then R is a minimum.

The technique is to rotate the system of axes until the diagonal terms in the matrix are made zero. Rotation is achieved by pre-multiplication and post-multiplication by a unit matrix [R] representing a rotation. The most reliable, although a somewhat slow way of finding this rotation is by using Jacobi's method, which is iterative. The largest off-diagonal term is reduced to zero by one rotation and so on successively. The values remaining along the diagonal are the eigenvalues, in this case the principal moments of inertia and the resultant of all the rotations applied gives the principal axes in terms of the initial system of axes. The program JACOBI is given in the appendix in both APL and BASIC.

An important use of the eigenvalue and vector program (JACOBI) is to find the best line or plane through a group of points in space.

The best line will be the axis for which the moment of inertia is a minimum, that is, the eigenvector corresponding to the smallest eigenvalue. The best plane will be that perpendicular to the eigenvector corresponding to the largest eigenvalue.

An elongated swarm of points will have an inertia tensor of reciprocal shape, namely an oblate ellipsoid, and a set of points nearly in a plane will have an inertia tensor which is nearly a line (perpendicular to that plane).

In many cases this calculation enables us to find the symmetry axis of an object or distribution, bearing in mind that the inertia tensor can show only the symmetry of an ellipsoid which may have three different axes, each perpendicular to a plane of symmetry.

Other uses of the eigenvalue program are to find the frequencies of the normal modes of vibration of a coupled system and, in statistics, as a general way of discovering correlations between variables in factor analysis.

3.8 The Solution of the Cubic Equation and the Diagonalisation of a 3 × 3 Real Symmetric Matrix

It is often necessary to find the eigenvalues of a symmetric 3 × 3 matrix, for example, to find the moments of inertia of a group of points to establish the best plane through them. These values can be found directly as the roots of a cubic equation, for the solution of which a direct formula is available, although Jacobi's iterative method for a 3 × 3 matrix is very fast. We can use the fact that such a matrix has three invariants, quantities which are unchanged by rotating the axial system.

The diagonal terms R_1, R_2, R_3 of the diagonalised matrix are the roots of the secular equation

$$\lambda^3 - T\lambda^2 + S\lambda - D = 0$$

where T, S, D are the invariants of the matrix, which are quantities derived from the matrix unchanged by rotation of the axes.

The initial matrix is

$$\begin{bmatrix} A & G & E \\ G & B & F \\ E & F & C \end{bmatrix}$$

and this must transform to

$$\begin{bmatrix} R_1 & 0 & 0 \\ 0 & R_2 & 0 \\ 0 & 0 & R_3 \end{bmatrix}$$

where

$T = A + B + C$ (the trace)

$S = AB - G^2 + AC - E^2 + BC - F^2$ (the second invariant)

$D = ABC + 2GEF - AF^2 - CG^2 - BE^2$ (the determinant).

The steps are

Find $P = T^2/3 + S$

$Q = -2T^3/27 + TS/3 - D$

$\alpha = \text{arc } \cos(-Q/(2\sqrt{(-(p/3)^3)}))$

then $R_1 = T/3 + 2\sqrt{(-P/3)} \cos(\alpha/3)$

$R_2 = T/3 - 2\sqrt{(-P/3)} \cos(\alpha/3 + 60°)$

$R_3 = T/3 - 2\sqrt{(-P/3)} \cos(\alpha/3 - 60°)$

Check that $R_1 + R_2 + R_3 = T$

$$R_1 R_2 + R_3 R_1 + R_2 R_3 = S$$

$$R_1 R_2 R_3 = D$$

The above procedure is used if the three solutions of the cubic equation are all real, as is always the case for the moments of inertia, but usually one root is real and two are complex conjugates. The procedure is then to substitute $x = y - a/3$ into the standard equation $x^3 + ax^2 + bx + c = 0$ giving $y^3 + py + q = 0$. Here $p = -a^2/3 + b$ and $q = 2a^3/27 - ab/3 + c$. The roots then are

$$y_1 = A + B$$

$$y_2 = (A + B)/2 + i(A - B)\sqrt{3}/2$$

$$y_3 = (A + B)/2 - i(A - B)\sqrt{3}/2$$

where $A = (-q/2 + \sqrt{Q})^{\frac{1}{3}}$ and $B = (-q/2 - \sqrt{Q})^{\frac{1}{3}}$ and the positive values of the cube roots are taken. These algorithms are readily transcribed into programs.

3.9 To Solve N Non-linear Equations $F_i(x_j) = 0$

The ability to solve systems of linear equations directly enables us to solve systems of non-linear equations by iterative methods. Non-linear equations occur in geometry in, for example, calculating distances.

We assume that there are more equations than unknowns x_j and that $N \geqslant M$.

There may be more than one solution and we must guess at an approximate solution and then hope to approach the solution which we want by an iterative process. Different starting points may give different solutions so that at the end we must check that we have the solution which we require. The procedure is as follows:

(1) Assume starting values of the M unknowns x_j.
(2) Calculate E_i, the discrepancies in the N equations $F_i(x_j) = 0$ when the values for x_j are substituted, that is $F_i(x_j) = E_i$. By minimising these discrepancies, we will approach a solution, which will be reached when all are sufficiently small.
(3) Calculate $\partial E_i/\partial x_j$, either *analytically* if the equations $F_i(x_j) = 0$ can be differentiated, or *numerically* by changing each value of x_j in turn by a small increment δx_j and finding the corresponding change δE_i in E_i.
(4) Invert the $(N \times M)$ array $(\partial E_i/\partial x_j)$ to $D(M \times N)$. The discrepancies in each equation are given by $E_i = (\partial E_i/\partial x_j) \delta x_j$. There are thus i equations each of j terms. These N linear equations must be solved for the changes δx_j in x_j which account for the errors. $\delta x_j = D_{ji} E_i$.
(5) Multiply E_i by D_{ji} to obtain the corrections δx_j.

(6) Apply $-\delta x_j$ to obtain improved values of x_j. If all the values of δx_j are less than a certain value then stop, but otherwise return to step (2) above.

(7) If the values of x_j do not correspond to the particular solution which we want, then the process must be begun again from a new starting point.

Note

(a) If we have N equations for N unknowns then the ordinary inversion procedure may be used. Otherwise either the generalised inverse or the least squares procedure is appropriate.

(b) In applying the corrections δx_j it is better to limit the amount of correction which may be applied at any one cycle. The function arctan (x) has a maximum of $\pi/2$ (and takes the sign of x), so that if we wish to limit corrections to a maximum of A, then instead of δx we should subtract A*2/PI*ATN(DX*PI*.5/A).

We have used this procedure to match the x, y, z coordinates of a set of points to a particular set of distances between the points and a program TELLUR is given in the appendix. Here, although it appears that $3N$ unknowns will be found for N points, given the $N(N-1)/2$ distances between them, in fact there are only $3N-6$ unknowns, since the orientation of the points as a whole with respect to the coordinate system is unspecified and the matrix which is to be inverted will be singular. In this case the generalised inverse will give the correct solution. This program answers the problem (first propounded by Lazare Carnot about 1812): given enough distances between members of a set of points to fix the configuration of the set, find any other distances between them (or any required bond or torsion angles).

3.10 Computation as a Substitute for Algebra

We may use computational methods as a substitute for algebra and take for example, the solution of a number of non-linear equations. We may take the expressions for the area S of a triangle of sides A, B, C:

$$S^2 = P(P-A)(P-B)(P-C)$$

where $P = 0.5(A+B+C)$ is the semi-perimeter. The radius R of the circumscribed circle is R and $R = ABC/(4S)$. Given S, P, R find A, B, C. The direct solution would involve awkward algebra to eliminate B and C and then the solution of a cubic equation for A, followed by back substitution to find B and C. A BASIC program SOLVE, to solve this problem is given in the appendix. Such a method may not always converge to the required solution and some trial in the starting values may be necessary.

4 Coordinate Systems

There are two ways of describing structures in space. (a) Looking from the outside, seeing the structure in an external framework of coordinate axes and assigning coordinates to each point. Various systems of coordinates are possible, Cartesian, cylindrical and spherical being the commonest. Algebraic relationships between the coordinates can be used to describe planes, lines, spheres and other geometrical forms. (b) Looking from the inside, seeing the immediate local surroundings of each point (or atom or molecule or whatever is taken as the unit) and describing the local surroundings.

Usually it is convenient to describe the geometry of a structure, such as a molecule, with respect to a coordinate system fixed in space. This means, if we are interested only in the internal geometry of the molecule, that the coordinates of each point have no individual physical significance. Moreover, the relationship of the molecule to the frame has to be specified, whether it is significant or not. Three coordinates must be specified for each point, making $3N$ for a molecule of N atoms, but only $(3N - 6)$ are needed to describe the molecular configuration in isolation. The extra six parameters give the position of the molecule as a whole with respect to the axial system.

4.1 Cartesian Coordinates

Ordinary Cartesian coordinates (called after René Descartes) x, y, z with respect to three mutually orthogonal axes require no further description. They are conventionally taken to be right-handed. Pythagoras's theorem supplies us with the distance d_{ij} between two points i and j as

$$d_{ij}^2 = (x_i - x_j)^2 + (y_i - y_j)^2 + (z_i - z_j)^2$$

Cartesian coordinates are simplest because there is no cross term (with products of x and y) in the expression for the distance.

4.2 Cylindrical and Spherical Coordinates

Cylindrical coordinates r, ϕ, z are useful when the system studied (such as a helix) has an axis of rotational symmetry. All other coordinate systems are referred to Cartesian coordinates for definition. The transformations are

$$x = r\cos\phi \qquad r = (x^2 + y^2)^{\frac{1}{2}} \qquad (r \geq 0)$$

$$y = r\sin\phi \qquad \phi = \arctan(y/x) \qquad (0 \leq \phi < 360°)$$

$$z = z$$

The element of volume equivalent to the Cartesian element dx dy dz is

$$r \, dr \, d\phi \, dz$$

Spherical coordinates are useful when there is a unique centre, such as in the geography of the Earth. Here (see figure 4.1)

$$x = r\sin\theta\cos\phi \quad r = (x^2 + y^2 + z^2)^{\frac{1}{2}} \; (r \geq 0)$$

$$y = r\sin\theta\sin\phi \quad \theta = \arctan((x^2 + y^2)^{\frac{1}{2}}/z) \; (0 \leq \theta < 180°)$$

$$z = r\cos\theta \quad \phi = \arctan(y/x) \; (0 \leq \phi < 360°)$$

The element of volume is here $r^2 \sin\theta \, dr \, d\theta \, d\phi$.

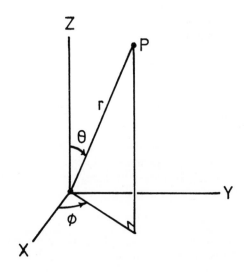

Figure 4.1. The relationship of polar coordinates (r, θ, ϕ) to the Cartesian coordinates (x, y, z) is $x = r\sin\theta\cos\phi, y = r\sin\theta\sin\phi, z = r\cos\theta$.

4.3 Area and Volume in Cartesian Coordinates

The volume V of a tetrahedron with vertices 1, 2, 3, 4 at positions x_1, y_1, z_1, etc. is given by the determinant

$$6V = \begin{bmatrix} x_1 & y_1 & z_1 & 1 \\ x_2 & y_2 & z_2 & 1 \\ x_3 & y_3 & z_3 & 1 \\ x_4 & y_4 & z_4 & 1 \end{bmatrix}$$

If the vertices have the sense indicated in figure 4.2 in relation to right-handed axes, then the volume has a positive sign. If the designations of any two vertices are exchanged then the sign of the volume is reversed. This expression is thus important for calculating on which side of the plane (123) the point (4) lies. Since the volume of a tetrahedron is $1/3 \times$ (area of base) $\times$ (height), the distance h_4 of the point (4) from the plane (123) is given by $V_{1234} = (1/3)A_{123}h_4$, where A_{123} is the positive area of the triangle (123).

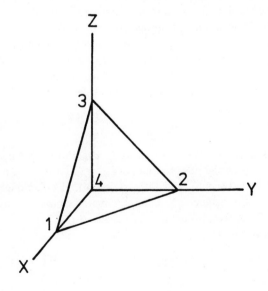

Figure 4.2. The sign of the volume of the tetrahedron 1, 2, 3, 4 is considered positive when the points are numbered as shown.

The area A_{123} of the triangle is best calculated from the distances d_{12}, d_{23}, d_{31} using Heron's formula

$$A_{123}^2 = s(s - d_{12})(s - d_{23})(s - d_{31})$$

where s is half the sum of the lengths of the sides.

4.4 Distances, Bond Angles and Torsion Angles

To build a structure, such as a molecule, of points joined by lines, three kinds of quantities are needed for describing the configuration in terms of the mutual positions of the points, that is, using internal coordinates. The usual measures are: d_{ij}, θ_{ijk} and ϕ_{ijkl} (distances, bond angles and torsion (or dihedral) angles). Of these only the first, distance, has the dimensions of length, the others being dimensionless ratios, although they may be related to distances, areas and volumes respectively.

Inter-point distances d_{ij}

Distances are readily calculated from the Cartesian coordinates of the points i and j as $d_{ij}^2 = (x_i - x_j)^2 + (y_i - y_j)^2 + (z_i - z_j)^2$, or they may be basic observational data.

Building up a framework, the second point lies at a distance d_{12} from the first. To add the third point two measures are needed, the distances d_{13} and d_{23} from the first and second points and for adding the fourth and subsequent points, three distances (or other parameters) must be given making $(3N - 6)$ for the whole assembly of N points. Distances are necessarily positive. Any set of distances between points must obey the triangle inequality — the sum of two sides of a triangle cannot be less than the third side.

Bond angles θ_{ijk}

A bond angle is given in terms of distances d_{ij}, d_{jk}, d_{kl} as $d_{ij}^2 + d_{jk}^2 - 2d_{ij}d_{jk}\cos\theta_{ijk}$ (the cosine rule) so that $\cos\theta_{ijk} = (d_{ij}^2 + d_{jk}^2 - d_{ik}^2)/(2d_{ij}d_{jk})$. The angle must be between $0°$ and $180°$ so that it is given uniquely by the cosine.

In terms of vector distances $\mathbf{d}_{ij}$ and $\mathbf{d}_{jk}$, directed from i to j and from j to k respectively, the scalar product $\mathbf{d}_{ij}.\mathbf{d}_{jk}$ equals $|\mathbf{d}_{ij}||\mathbf{d}_{jk}|(-\cos\theta_{ijk})$. So that, using Cartesian coordinates

$$-\cos\theta_{ijk} = \frac{(x_j - x_i)(x_k - x_j) + (y_j - y_i)(y_k - y_j) + (z_j - z_i)(z_k - z_j)}{|d_{ij}||d_{jk}|}$$

where

$$|d_{ij}| = ((x_j - x_i)^2 + (y_j - y_i)^2 + (z_j - z_i)^2)^{\frac{1}{2}} \text{ etc.}$$

Similarly, the vector product $d_{ij} \times d_{jk}$ is a vector perpendicular to the plane of the two vectors and of magnitude proportional to the area of the parallelogram formed by the two vectors which is $|d_{ij}||d_{jk}| \sin \theta_{ijk}$. Thus

$$\sin \theta_{ijk} = \frac{|d_{ij} \times d_{jk}|}{|d_{ij}||d_{jk}|}$$

Since θ is defined as being between 0 and $180°$ the sign of its sine is not given by the above expression but must be allocated by assessing the handedness of the three vectors d_{ij}, d_{jk} and their vector product. The hand is shown by figure 4.3. For calculating θ_{ijk} the cosine expression is clearly preferable.

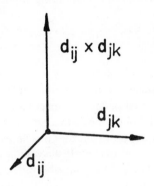

Figure 4.3. The sign of the vector product.

Tetrahedral bonds

When four straight lines meet at a point, six bond angles are formed between them. However, only five quantities are required to specify the configuration and thus there is a relationship between the six angles which enables any one to be found in terms of the other five as the solution of a quadratic equation

$$\begin{vmatrix} 1 & \cos\theta_{12} & \cos\theta_{13} & \cos\theta_{14} \\ \cos\theta_{12} & 1 & \cos\theta_{23} & \cos\theta_{24} \\ \cos\theta_{13} & \cos\theta_{23} & 1 & \cos\theta_{34} \\ \cos\theta_{14} & \cos\theta_{24} & \cos\theta_{34} & 1 \end{vmatrix} = 0$$

It derives from the consideration that the four axes define a four-dimensional unit cell and that, since we are working in three-dimensional space, this four-dimensional content should be zero. This determinant is the square of that content and thus zero.

If more than four lines, for example N lines, meet at a point then the matrix

$$
\begin{bmatrix}
1 & \cos\theta_{12} \ldots & \cos\theta_{1N} \\
\cos\theta_{12} & & \cdot \\
\cdot & & \cdot \\
\cdot & & \cdot \\
\cdot & & \cdot \\
\cos\theta_{1n} & & 1
\end{bmatrix}
$$

should be of rank three and all determinants of order four made from it should be zero. In other words the four-line relationship applies to all combinations of four lines among the N.

Torsion angles

Three points 1,2,3 in a triangle specify a bond angle θ_{123}. When a fourth point (4) is to be added, three further parameters must be specified. These may be distances from the points 1,2,3 or one distance, such as d_{34}, one bond angle θ_{234} and one torsion angle ϕ_{1234}. In all cases it is necessary to know on which side of the plane (123) the point (4) lies. If the volume of the tetrahedron (1234) is calculated from the x, y, z coordinates of the four points, then its sign will be positive if (4) lies on one side of (123) and negative if it lies on the other. The sign requires that the hand of the axial frame should be fixed, so that right-handed axes are always chosen.

The torsion angle ϕ_{1234} (which equals ϕ_{4321}) is defined as the dihedral angle between the planes (123) and (234) and is positive in the case illustrated in figure 4.4. The ϕ value for the *cis*-configuration is $0°$ and for the *trans*-configuration is $180°$. ϕ may run from $0°$ to $360°$ but it is usually better to make it lie between $-180°$ and $+180°$ since mirror symmetry is then more recognisable. Looking at the tetrahedron of figure 4.2 the angle ϕ_{1234} is seen to be negative while the volume of the tetrahedron (as given by the formula of section 4.3) is positive so that V_{1234} and ϕ_{1234} have opposite signs.

ϕ_{1234} is the angle between the normals to the planes (123) and (234) which are parallel to the vector products $(d_{12} \times d_{23})$ and $(d_{23} \times d_{34})$ respectively. For the coordinate axes $X \times Y = Z$. Thus

$$
\begin{aligned}
\cos\phi_{1234} &= \frac{(d_{12} \times d_{23}) \cdot (d_{23} \times d_{34})}{|d_{12} \times d_{23}||d_{23} \times d_{34}|} \\[2mm]
&= \frac{\begin{vmatrix} d_{12} \cdot d_{23} & d_{23} \cdot d_{23} \\ d_{12} \cdot d_{34} & d_{23} \cdot d_{34} \end{vmatrix}^{\frac{1}{2}}}{\begin{vmatrix} d_{12}^2 & d_{12} \cdot d_{23} \\ d_{12} \cdot d_{23}^{23} & d_{23} \end{vmatrix}^{\frac{1}{2}} \begin{vmatrix} d_{23}^2 & d_{23} \cdot d_{34} \\ d_{23} \cdot d_{34} & d_{34}^2 \end{vmatrix}^{\frac{1}{2}}}
\end{aligned}
$$

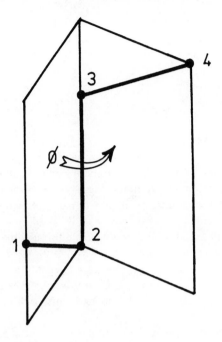

Figure 4.4. The sign of the torsion angle ϕ_{1234} is positive.

$$= \frac{\cos\theta_{123} \cdot \cos\theta_{234} - \cos\theta_{(1234)}}{\sin\theta_{123} \cdot \sin\theta_{234}}$$

(By cancelling $d_{12}d_{33}d_{23}^2$ and using $\boldsymbol{d}_{12} \cdot \boldsymbol{d}_{23} = -d_{12}d_{23}\cos\theta_{123}$

$$\cos\theta_{(1234)} = \left(\left|\frac{\boldsymbol{d}_{12} \cdot \boldsymbol{d}_{34}}{|\boldsymbol{d}_{12}||\boldsymbol{d}_{34}|}\right|\right)$$

These scalar and vector products are readily calculated from the orthonormal Cartesian coordinates. For example

$$\boldsymbol{d}_{12} \cdot \boldsymbol{d}_{34} = (x_2 - x_1)(x_4 - x_3) + (y_2 - y_1)(y_4 - y_3) + (z_2 - z_1)(z_4 - z_3)$$

$$|\boldsymbol{d}_{12}| = [(x_2 - x_1)^2 + (y_2 - y_1)^2 + (z_2 - z_1)^2]^{\frac{1}{2}}$$

but $\cos\phi_{1234}$ can also be expressed in terms of internal parameters as

$$\cos\phi_{1234} = \frac{d_{13}^2 - d_{14}^2 + d_{24}^2 - d_{23}^2 + 2d_{12}d_{34}\cos\theta_{123}\cos\theta_{234}}{2d_{12}d_{34}\sin\theta_{123}\sin_{234}}$$

When the inverse cosine is taken the resulting angle is between $0°$ and $180°$. The correct sign must then be attached to accord with the hand of the configuration (1234). The hand of the torsion angle (shown in figure 4.4) is positive, the opposite to the sign of the volume as determined from the formula above. The sign of the volume of a tetrahedron can be found by inspection on comparing the configuration with that of the tetrahedron in figure 4.2, which is positive.

4.5 Lines and Planes

For the purpose of plotting points on a screen, it is usually best to introduce equations in the form of parametric equations, so that for a line the continuous variation of one parameter causes the line to be traced out and for a surface, scanning two parameters over appropriate ranges generates the required segment of surface (as a network of points, which may or may not be resolvable from each other).

Thus, the parametric equations generating a line through the point x_1, y_1, z_1, with the direction cosines l, m, n, are $x = x_1 + \lambda l, y = y_1 + \lambda m, z = z_1 + \lambda n$.

To draw a line from x_1, y_1, z_1 to x_2, y_2, z_2 we generate $x = x_1(1 - \lambda) + x_2 \lambda$, $y = y_1(1 - \lambda) + y_2 \lambda, z = z_1(1 - \lambda) + z_2 \lambda$. Clearly $0 < \lambda < 1$ will give the segment required but the line can be produced by taking λ outside this range. Usually only a limited segment of a line is required.

Frequent problems are: to draw the 'best' line or the 'best' plane through a cluster of points in space and to find the deviations of the points from this line or plane. The usual criterion for the best fit is that the sum of the squares of the deviations of the points from the line or plane should be a minimum (the least squares criterion). For the best line the moment of inertia of the system of points (about the line) should be a minimum. For the best plane the normal is the direction of greatest moment of inertia. The procedures involve, then, finding the tensor of inertia and diagonalising it. The centre of gravity is the appropriate origin when calculating the moment of inertia.

4.6 To Draw an Ellipse

The equation of an ellipse with its centre at the origin and with semi-axes a and b is $(x/a)^2 + (y/b)^2 = 1$ and this can be expressed in terms of a parameter t which runs from 0 to 2π as: $x = a \cos(t)$ and $y = b \sin(t)$. If the centre of the ellipse is at x_1, y_1, then $x = x_1 + a \cos(t)$ and $y = y_1 + b \sin(t)$, and if the phase of the ellipse is to be changed by a clockwise rotation through the angle α then

$$x = a \cos(\alpha).\cos(t) + b \sin(\alpha).\sin(t)$$

$$y = -a \sin(\alpha).\cos(t) + b \cos(\alpha).\sin(t)$$

Many microcomputers have circle and line drawing as supplied functions.

4.7 The Helix

The parametric equations for a helix (in terms of the parameter t) are

$$x = a \cos(t*(a^2 + b^2)^{\frac{1}{2}})$$
$$y = a \sin(t*(a^2 + b^2)^{\frac{1}{2}})$$
$$z = bt/(a^2 + b^2)^{\frac{1}{2}}$$

4.8 The Ellipsoid

The parametric equations (with two parameters u and v) are (deriving them from spherical coordinates): $x = a \cos(u) \sin(v)$, $y = b \sin(u) \sin(v)$, $z = c \cos(v)$ so that $(x/a)^2 + (y/b)^2 + (z/c)^2 = 1$, the equation of an ellipsoid centred at the origin with its axes in the directions of the coordinate axes.

4.9 Rotations

Coordinates x, y, z can be rotated to x', y', z' by a rotation matrix $[R]$. Repeated rotations are produced by successive multiplications $[X'] = [X] [R^{(n)}] \ldots [R^{(1)}]$. (It is convenient to make $[X]$ a row matrix with dimensions $<1,3>$ so that $X'<1,3>=X<1,3>*R<3,3>$, so that to post-multiply to give the new coordinates, but other conventions are possible.) This is called 'concatenation'. It may happen that particular terms in $[X]$ become very small so that, when they are again multiplied up, accuracy is lost. For this reason and for general convenience, it is usual to use homogeneous coordinates $(x, y, z, w,)$ to denote a single point. The actual coordinates along the three Cartesian axes are $x/w, y/w, z/w$ where w is not equal to 0. However, for the present purposes it is sufficient to make $w = 1$ and use a matrix $[X] = [x, y, z, 1]$.

A 3×3 rotation matrix $[R]$ for rotation by an angle θ (clockwise looking into the origin) about the Z-axis (which has the direction cosines $0, 0, 1$) takes the form

$$[R] = \begin{bmatrix} \cos\theta & -\sin\theta & 0 \\ \sin\theta & \cos\theta & 0 \\ 0 & 0 & \end{bmatrix}$$

And more generally, for a rotation by an angle θ about the line with direction cosines l, m, n

$$[R] = \begin{bmatrix} ll(1-\cos\theta)+\cos\theta, & ml(1-\cos\theta)-n\sin\theta, & nl(1-\cos\theta)+m\sin\theta \\ lm(1-\cos\theta)+n\sin\theta, & mm(1-\cos\theta)+\cos\theta, & nm(1-\cos\theta)-l\sin\theta \\ ln(1-\cos\theta)-m\sin\theta, & mn(1-\cos\theta)+l\sin\theta, & nn(1-\cos\theta)+\cos\theta \end{bmatrix}$$

This matrix has a unit determinant and depends on only three variables (since $l^2+m^2+n^2=1$).

When using homogeneous coordinates of the form $[X]=[x,y,z,1]$ this 3×3 rotation matrix can be embedded in a 4×4 matrix to give a matrix which permits a variety of other transformations. The full 4×4 matrix is

$$\begin{bmatrix} R_{11} & R_{12} & R_{13} & 0 \\ R_{21} & R_{22} & R_{23} & 0 \\ R_{31} & R_{32} & R_{33} & 0 \\ T_1 & T_2 & T_3 & S \end{bmatrix}$$

Here the additional terms T_1 represent a translation and the final term S represents a scale. S multiplies the last term w and thus if $S=2$ the three coordinates are divided by 2 and the whole object is halved in size.

Perspective transformations may also be carried out.

4.10 Crystallographic and Orthonormal Axes

An important application of solid geometry is to the shapes of molecules, and data about the dimensions of molecules is normally provided in terms of the fractional coordinates of the atoms with respect to the crystallographic axes of the unit cell. These axes are frequently oblique and usually of different lengths and so, before distances and angles in molecules can be calculated, the atomic positions must be referred to the usual Cartesian coordinate system. The axial system may be oblique with steps of a, b and c along the three axes and inter-axial angles α (*alpha*, between b and c), β (*beta* between a and c) and γ (*gamma*, between a and b).

The crystal faces and the diffraction effects of a crystal are related to another set of oblique axes, the *reciprocal axes* (which are denoted by marking with an asterisk) and it is convenient to give all the necessary transformations at the same time. Such reciprocal axes are the basis of Fourier space (see chapter 10).

Orthonormal coordinates X(1), X(2), X(3) are obtained from the fractional coordinates XC(1), XC(2), XC(3) by multiplication by a 3×3 array L (see figure 4.5)

X(3) = L(3,1)*XC(1) + L(3,2)*XC(2) + L(3,3)*XC(3)

X(2) = L(2,1)*XC(1) + L(2,2)*XC(2)

X(1) = L(1,1)*XC(1)

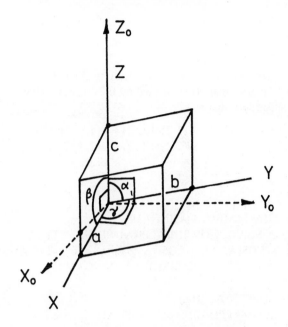

Figure 4.5. The conversion of oblique crystallographic axes X, Y, Z, to the orthonormal (Cartesian) axes X_0, Y_0, Z_0. It is convenient to take Z_0 parallel to Z and X_0 in the X-Z plane. The coordinates of atoms in the unit cell of a crystal are given as fractions of the unit cell sides, a, b, c, which make angles α, β, γ with each other instead of $90°$.

If the multiplication is done in this order the array XC can be overwritten by the new array X. Such stratagems were important for the first computers where storage space was extremely restricted (and a computer weighing a ton would have a storage space at 4K bytes). (Here the angles between the reciprocal axes are named ALPHAST, BETAST and GAMST.)

The array L is first calculated as follows

COSGAMST = (COS(ALPHA)*COS(BETAST) − COS(GAMMA))/
(SIN(ALPHA)*SIN(BETAST))

SINGAMST = SQR(1 − COSGAMST*COSGAMST)

    L(1,1) = A*SIN(BETA)*SINGAMST
    L(1,2) = 0
    L(1,3) = 0
    L(2,1) = − A*SIN(BETA)*COSGAMST
    L(2,2) = B*SIN(ALPHA)
    L(2,3) = 0

$$L(3,1) = A*COS(BETA)$$
$$L(3,2) = B*COS(ALPHA)$$
$$L(3,3) = C$$

$[X] = [L] [XC]$ and the reverse transformation is $[XC]=[L^{-1}] [X]$.

The diffraction effects are transformed by the corresponding transformation of reciprocal space coordinates

$$H(1) = U(1,1)*HC(1) + U(1,2)*HC(2) + U(1,3)*HC(3)$$
$$H(2) = \qquad\qquad U(2,2)*HC(2) + U(2,3)*HC(3)$$
$$H(3) = \qquad\qquad\qquad\qquad\qquad U(3,3)*HC(3)$$

where

$$U(1,1) = 1/(A*SIN(BETA)*SINGAMST)$$
$$U(1,2) = COSGAMST/(B*SIN(ALPHA)*SINGAMST)$$
$$U(1,3) = -(COS(ALPHA)*COSGAMST)/(C**SIN(ALPHA)*SINGAMST)$$
$$\qquad\qquad -COS(BETA)/(C*SIN(BETA)*SINGAMST)$$
$$U(2,1) = 0$$
$$U(2,2) = 1/(B*SIN(ALPHA))$$
$$U(2,3) = -COS(ALPHA)/(C*SIN(ALPHA))$$
$$U(3,1) = 0$$
$$U(3,2) = 0$$
$$U(3,3) = 1/C$$

L is a lower triangular matrix and U an upper. $[U] = [L^T]^{-1}$.

We require the orthonormal coordinates of real and reciprocal systems to correspond, which they do for the transformations given. Real and reciprocal spaces are related as follows. If $[G]$ is the metric matrix for real space and

$$[G] = \begin{bmatrix} a.a & a.b & a.c \\ b.a & b.b & b.c \\ c.a & c.b & c.c \end{bmatrix}$$

the metric matrix $[G*]$ for the reciprocal space is

$$\begin{bmatrix} a*.a* & a*.b* & a*.c* \\ b*.a* & b*.b* & b*.c* \\ c*.a* & c*.b* & c*.c* \end{bmatrix}$$

and $[G*]$ is the inverse of $[G]$. Each can be obtained as the inverse of the other and the individual elements can then be derived. A standard reference is Rollett (1965).

4.11 Networks

Ohm's Law

Geometry lies at the basis of networks and we will first take an example of the solution of a network, which might be simply to find the current in each branch of the network and the voltage at each node, when a voltage is applied between two points. Figure 4.6a shows a framework of 8 nodes connected by 12 branches. Since the figure is that of a cube, which can be mapped on to the surface of a sphere, it is clear how to identify the necessary number of meshes. (A network in the plane or on a sphere has a dual, where nodes replace faces and vice versa.) We have to find 12 currents and have 6 mesh equations and 8 node equations.

We begin by marking a direction on each branch (for the moment we assume that each branch is of unit resistance). The mesh equations, Kirchhof's voltage equations, express the fact that the voltages summed round a closed loop add to zero. Here mesh currents are counted positive when clockwise looked at from outside the cube. The voltages are the products of the current in each branch and its resistance. The node equations express the conservation of current. The sum of the currents entering and leaving any node must be zero. These are Kirchhof's current equations.

There are B branches, N nodes and M meshes. We have the NODE/BRANCH matrix

$A(n, b)$ = +1 if the branch b is directed out of node n
　　　 = −1 if the branch b is directed into node n
　　　 = 0 if the branch b is not incident on node n

and the BRANCH/MESH matrix

$M(m, b)$ = +1 if branch b bounds mesh m in the positive direction
　　　 = −1 if branch b bounds mesh m in the negative direction
　　　 = 0 if branch b does not belong to mesh m

$J(i)$ is the vector of all branch currents.
$V(i)$ is the vector of all branch voltages. It is: $[0\ 0\ 0\ 0\ 0\ 0\ 0\ 0\ 0\ 0\ 100\ -100]$.
(A voltage of 100 is applied across the corners of a cubic framework.)

Incidence matrix of branches on meshes: M(m, b)

	(12	23	34	41	26	37	48	15	56	67	58	78)
	1	2	3	4	5	6	7	8	9	10	11	12
(1234)	1	1	1	-1	0	0	0	0	0	0	0	0
(2367)	0	-1	0	0	1	-1	0	0	0	0	0	0
(1485)	0	0	0	1	0	0	1	-1	0	0	-1	0
(3784)	0	0	-1	0	0	1	-1	0	0	0	0	1
(5678)	0	0	0	0	0	0	0	0	-1	-1	1	-1
(1265)	-1	0	0	0	-1	0	0	1	1	0	0	0

This is of rank $(N - 1) = 5$.

Incidence matrix of branches on nodes: A(n, b)

1	1	0	0	1	0	0	0	1	0	0	0	0
2	-1	1	0	0	1	0	0	0	0	0	0	0
3	0	-1	1	0	0	1	0	0	0	0	0	0
4	0	0	-1	-1	0	0	1	0	0	0	0	0
5	0	0	0	0	0	0	0	-1	1	0	1	0
6	0	0	0	0	-1	0	0	0	-1	1	0	0
7	0	0	0	0	0	-1	0	0	0	-1	0	1
8	0	0	0	0	0	0	-1	0	0	0	-1	-1

This matrix is of rank 7. $MA^T = 0$ and $AM^T = 0$.
We have

$$M<6,12> \ I<12,1> = V<6,1>$$

and

$$R<8,12> \ I<12,1> = V<8,1>$$

these add to give

$$R<14,12> \ I<12,1> = V<14,1>$$

We solve $R<14,12>*I<12,1>=V<14,1>$ for I. R is not square so that we use either the least squares or the generalised inverse method. This gives I(k), the mesh/current vector, as (figure 4.6b)

1	-8.3333
2	-4.1667
3	16.6667
4	4.1667
5	-4.1667
6	-20.8333
7	20.8333
8	4.1667

9	−16.6667
10	−20.8333
11	20.8333
12	58.3333

$A(i, j) J(i) = 0$ is Kirchhof's current law — for each node j the sum of the currents i is zero. $M(i, k) V(i) = 0$ for each mesh k — the voltages in the branches round the mesh add to zero. From the topology of the system we have $M^T(k, i) A(i, j) = 0$ and $A^T(j, i) M(i, k) = 0$.

A network which has no dual

Kuratowski showed that the condition that a graph should have a dual, this is, that is should be capable of being wrapped over a spherical surface without overlapping branches, is that it should contain no sub-graphs of the types shown in figure 4.6c. The first of these K_1 is the complete graph on 5 points, the four-dimensional simplex, and the second K_2 is the basic of the well-known 'gas, water, electricity problem' (supply gas, water and electricity from three stations to three houses without allowing the pipes or wires to cross each other). Such graphs are not equivalent to polyhedra and we are unable to select the right number of mesh equations easily. However, we may simply take 'enough' mesh equations and need not worry whether some are linear functions of others.

For the graph K_1 the node/branch incidence matrix $A(n, b)$ is

Branch		12	13	14	15	23	24	25	34	35	45
Node	1	1	1	1	1	0	0	0	0	0	0
	2	−1	0	0	0	1	1	1	0	0	0
	3	0	−1	0	0	−1	0	0	1	1	0
	4	0	0	−1	0	0	−1	0	−1	0	1
	5	0	0	0	−1	0	0	−1	0	−1	−1

It is of rank $(N - 1)$ which is 4.

The mesh/branch incidence matrix $M(m, b)$ is (figure 4.6d)

		12	13	14	15	23	24	25	34	35	45
(123)	1	1	0	0	0	1	0	0	0	0	0
(124)	2	1	0	−1	0	0	1	0	0	0	0
(125)	3	1	0	0	−1	0	0	1	0	0	0
(134)	4	0	1	−1	0	0	0	0	1	0	0
(135)	5	0	1	0	−1	0	0	0	0	1	0
(145)	6	0	0	1	−1	0	0	0	0	0	1
(234)	7	0	0	0	0	1	−1	0	1	0	0
(235)	8	0	0	0	0	1	0	−1	0	1	0
(245)	9	0	0	0	0	0	1	−1	0	0	1
(345)	10	0	0	0	0	0	0	0	1	−1	1

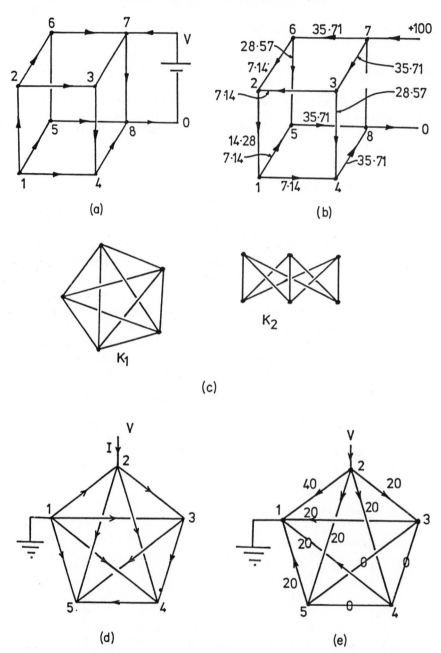

Figure 4.6. A cubic spatial network of resistors forming branches and loops (meshes). Here, since the cube can be mapped on to the two-dimensional surface of a sphere, counting the meshes may be unambiguous.

This is of rank 6.

The matrices $AM^T=0$ and $MA^T=0$ from topological considerations and this forms a useful check against copying mistakes.

Suppose that a potential V is applied at (2) and point (1) is earthed and that this produces a current of 100. The vector of the right-hand sides of the equations is then: r.h.s. = [−100 100 0 0 0] for the node/branch block of equations (Kirchhof's current law) and [0 0 0 0 0 0 0 0 0 0 0 0] for the mesh equations (Kirchhof's voltage law).

Solving these 15 equations for the 18 currents gives

1	2	3	4	5	6	7	8	9	10
−40,	−20,	−20,	−20,	20,	20,	20,	0,	0,	0 (figure 4.6e)

If we apply a voltage of 80 across (21) then the current in (21) (for unit conductivity) will be 80, instead of 40, so that all the currents above must be doubled.

The total current flowing out of node 2 will be then 80+40+40 (21)+(23)+ (24)+(25) so that the source must supply a current of 160 for this voltage of 80 so that the effective resistance of the network between (1) and (2) is 80/160=0.5.

If the conductivities of the various branches are different, then the matrix A of connections can be multiplied by the matrix of conductivities.

A relevant reference is Oster and Desoer (1971).

5   Geometry on the sphere: spherical trigonometry

Geometry on the surface of a sphere is of importance for geography and for navigation, particularly in the era of satellites, but is also valuable for the many questions of directions radiating from a point which can be represented as positions on a sphere. In spherical astronomy, where distances may be undetermined, the positions of stars and planets are represented on a celestial sphere. On the surface of a sphere we see a new kind of geometry where measurement is different from that of the plane and this forms an introduction to more complex types of geometry.

Spherical trigonometry, the basis of most calculations on a sphere, has been revolutionised, earlier by the provision of logarithm tables, then by the simple calculating machine, and later by the advent of the programmed computer. Solving spherical triangles using old-fashioned seven-figure logarithm tables was a penance, sometimes compounded by sea-sickness, from which calculators have relieved us.

Compare solving: $\cos(a) = \cos(b)\cos(c) + \sin(a)\sin(b)\cos(c)$ with seven-figure trigonometric tables and with a simple electronic calculator! Such an equation may now be solved thousands of times, as part of an iterative process, with far less effort.

Distances are measured along the surface of a sphere and are reckoned as angles subtended at the centre, either degrees or radians, and linear distance is obtained by multiplying by the radius R of the sphere. The shortest distance between two points (a geodesic) is the arc of a circle, the centre of which coincides with the centre of the sphere. The plane containing a great circle passes through the centre of the sphere. Such circles on a sphere, which have the same radius as that of the sphere, are called *great circles*. An elastic band stretched between the two points would follow this great circle course. The other part of the great circle is the longest distance between the two points. Any other circle on the sphere is called a *small circle*. On the Earth, the Equator and all the meridians passing through the North and South Poles are great circles. The circles of constant latitude, such as the Tropics of Cancer and Capricorn, are small circles. A circle represents the locus of points at a given distance from its centre (figure 5.1a).

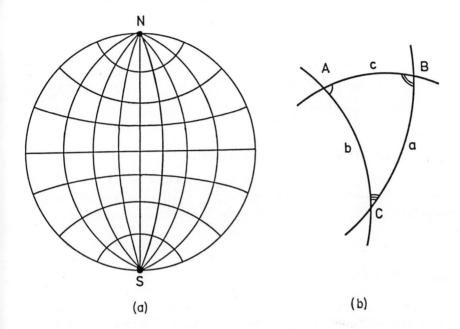

(a)   (b)

Figure 5.1. (a) On a sphere *great circles* are circles on the surface which have
the same radius as that of the sphere. The meridians of longitude are great
circles. Circles of smaller radius, such as the parallels of latitude, are *small
circles*. (b) A spherical triangle is one bounded by arcs of three great circles.
The great circle track is the shortest distance (geodesic) between two points.

     The basic element is the *spherical triangle* (figure 5.1b) which is bounded
by arcs of three *great circles* joining the three points A, B and C. A spherical
triangle thus has three lengths a, b and c for its sides BC, AC and AB respectively and has three angles, A, B and C, at its vertices. These angles A, B, C
may be visualised in three ways: (a) the angle A at the vertex A is the angle
between the tangents at A to the great circles AB and AC (the tangent lines
are also tangents to the sphere at A); (b) the angle A is also the dihedral angle
at which the planes of the two great circles AB and AC intersect (at the centre
of the sphere); and (c) it is also the angle between the normals to these meridian
planes. The normal to the plane of a great circle may be drawn from the centre
of the sphere to intersect its surface (in two diametrically opposite points).
These points may be referred to as the poles of the great circle. Thus, the North
and South Poles on the Earth are the poles of the equatorial plane. Solving a
spherical triangle consists, as in solving a plane triangle, in finding all the six
elements A, B, C, a, b, c when enough data is given. Usually a knowledge of
three elements will suffice, but there are special cases.

The fundamental formulae of spherical trigonometry are as follows. The cosine rule (to find a side)

$$\cos(a) = \cos(b)\cos(c) + \sin(b)\sin(c)\cos(A)$$

The cosine rule (to find an angle)

$$\cos(A) = -\cos(B)\cos(C) + \sin(B)\sin(C)\cos(a) \text{ [note the minus sign]}$$

The sine rule

$$\sin(a)/\sin(A) = \sin(b)/\sin(C) = \sin(c)/\sin(C)$$

There are dozens of other formulae, but we may mention only two classes: the procedure for solving a triangle where one element, side or angle, is 90°, and the calculation of area.

5.1 Napier's Rules

The procedure, devised by Napier, the inventor of logarithms, is widely used where one element of a spherical triangle is a right-angle. We move round the triangle writing down the sides and angles, beginning with the right-angle, known or unknown in a diagram with five compartments. The three compartments opposite the right-angle must first be marked with '90° – '. Each compartment has two elements opposite to it and two elements adjacent. The sine and cosine rules above then reduce to the mnemonic (figure 5.2)

The *sine* of an angle is equal to the product of the *tangents* of the two angles *adjacent* to it and to the product of the *cosines* of the two angles *opposite* to it in the diagram.

5.2 The Area of a Triangle

The area of a spherical triangle is equal to its spherical excess, which is the amount by which the sum of the internal angles of the triangle exceed 180° or 2π. The area of the whole spherical surface is thus 720° or 4π in angular measure, where the units are properly called steradians rather than radians, but are without dimension. The actual area of the surface is found by multiplying the area in steradians by R^2. Thus, the area S is $(A + B + C - 2\pi)$ in terms of the angles.

In terms of the lengths of the sides we may use a formula due to Euler

$$\cos(S/2) = \frac{[\cos^2(a/2) + \cos^2(b/2) + \cos^2(c/2) - 1]}{[2\cos(a/2)\cos(b/2)\cos(c/2)]}$$

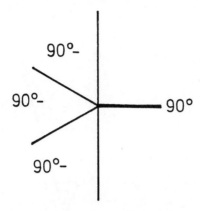

Figure 5.2. In a spherical triangle (such as that of figure 5.1b), if one angle or side is $90°$, the sine and cosine formula for a general triangle simplify to give Napier's Rules. The six elements (3 sides and 3 angles alternately in cyclic order) are entered in turn in this diagram with the right-angle element as shown and we have then the relationships: sine middle part = product of the tangents of adjacent parts and sine middle part = product of cosines of opposite parts.

5.3 Stereographic Projection

The problem of mapping the spherical vault of the stars and the spherical surface of the Earth on the plane has been with us ever since the Earth was conceived of as being round. Very many map projections are known and we will mention only three. *Stereographic projection* was known to Ptolemy of Alexandria in the second century AD. The method is very useful for representing directions radiating from a point as well as positions on a spherical surface. Both can be represented by latitude and longitude. The usual spherical coordinates r, θ and ϕ are respectively the radius (in this case constant) the co-latitude (angular distance from the North Pole) and the longitude (from an arbitrary meridian).

To represent a point in the Northern hemisphere of the sphere on a plane, usually taken as the plane of the Equator, a line is drawn from the South Pole to the point in question and the intersection of this line with the equatorial plane is the stereographic projection of the point. Points in the Southern hemisphere are conveniently represented in a separate diagram so that all points fall into the two circles. However, points in the Southern hemisphere may also be represented by joining them to the South Pole, in which case they project outside the circle of the equator. Thus, points near the South Pole project at great distances from the centre which is inconvenient (figure 5.3).

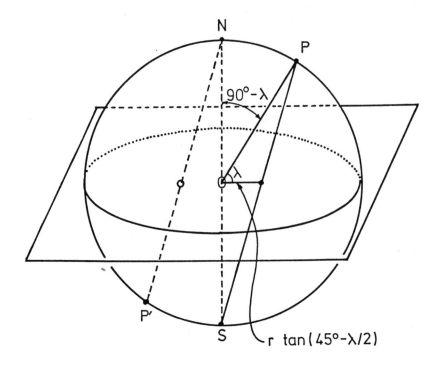

Figure 5.3. The stereographic projection of points on the surface of a sphere on to the equatorial plane is made by joining points in the Northerm hemisphere to the South Pole, those in the Southern hemisphere to the North Pole.

The important properties of the stereographic projection are: that both great circles and small circles project as circles (although their centres do not project to become the centres of the projected circles); angles between great circles are preserved — the projection is angle true; and areas are not preserved. The projection can thus be drawn with ruler and compasses and there is a large body of traditional know-how available on drawing and measuring such projections. For our purposes two disc diagrams representing Northern and Southern hemispheres separately are convenient for keeping visual track of angular relationships in space.

If we project to the equatorial plane, a point at latitude λ and co-latitude $(90 - \lambda)^{\circ}$ is projected to a distance $r \tan(90 - \lambda)/2$ from the centre in a direction which corresponds directly to the longitude.

Coordinates may be transformed if we wish to project in another direction. Having drawn a projection of the points we can see what spherical triangles need to be solved to calculate what we want. For example, the distance PQ (in angular measure) between two points P and Q whose latitudes (λ_P, λ_Q) and longitudes (L_P, L_Q) are known is obtained from the cosine rule

$$\cos(PQ) = \cos(90 - \lambda_P)\cos(90 - \lambda_Q) + \sin(90 - \lambda_P)\sin(90 - \lambda_Q)\cos(L_P - L_Q)$$

The angle NPQ, giving the direction in which, for example, a radio aerial must be pointed to send signals from P to Q, may be obtained from the sine rule

$$\sin(NPQ)/\sin(90 - \lambda_Q) = \sin(PNQ)/\sin(PQ)$$

See figure 5.4.

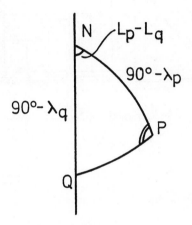

Figure 5.4. The great circle distance between two points P and Q, with latitudes λ_P and λ_Q and longitudes L_P and L_Q, on a sphere is found by the cosine rule, and then the angle NPQ by applying the sine rule.

5.4 Mercator's Projection

Here the sphere is projected (figure 5.5) from its centre on to a cylinder with its axis parallel to the N–S axis of the Earth. Points at high latitudes thus appear very far from the Equator, but as this projection is mainly used for marine navigation, and high latitudes are inaccessible, this fault was tolerated. The longitude is marked out linearly and the latitude coordinate y is proportional to the tangent of the co-latitude.

The key property which made Mercator's projection useful is that if a sailor wishes to sail from P to Q with a constant compass bearing then that bearing can be read from the chart. The course followed is a rhumb line and not a great circle course and is consequently longer than necessary.

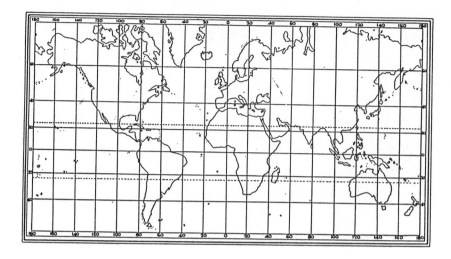

Figure 5.5. Mercator's projection of the sphere. The area becomes badly distorted near the poles.

5.5 The Hammer–Aitoff Projection

This is sometimes a convenient way to display the whole globe (figure 5.6). Recently it has been used to show features of the surface of protein molecules. One hemisphere of the Earth, 'the front', is projected into a circle and the 'back' hemisphere is put to left and right of this to fill up an ellipse of axes $2\sqrt{2}$ and $\sqrt{2}$ (and thus of area 4π). The parallels and meridians cut at angles not very far from right-angles and so shape is fairly well represented. Area is correctly represented. Coordinates (x, y) in the map of a point with latitude λ and longitude L are

$$x = [2\sqrt{2}\sin(90-\lambda)\sin((L-L_0)/2)]/[1+\sin(90-\lambda)\cos((L-L_0)/2)]^{\frac{1}{2}}$$
$$y = \sqrt{2}/[1+\sin(90-\lambda)\cos((L-L_0)/2)]^{\frac{1}{2}}$$

L_0 is the latitude allocated to the centre point in the projection.

5.6 Optimisation

With modern computing facilities many problems can be tackled by 'brute force', that is by allowing the computer to search in multi-dimensional parameter space for a maximum. There are several techniques for integrating over many variables. We may take a typical mathematical problem for which there

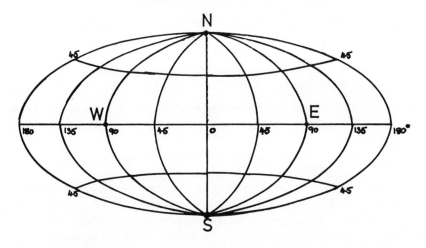

Figure 5.6. The Hammer–Aitoff projection of the sphere. The rear part of the sphere appears at the sides. This is an equal area projection and local shape is fairly well preserved.

is no 'mathematical' solution. The problem is easily stated but is difficult to solve. It is

> To pack *N* points on to the surface of a sphere so that they should be the maximum distance apart — so that the minimum distance between any two should be a maximum.

The answer is known for quite a large number of values of *N*. They are mostly rather symmetrical arrangements. The characteristic of a solution must be that it should be mechanically stable and if we consider the points as the centres of equal discs touching each other, no movement of any disc, or group of discs, with respect to the others should be possible, although there may, in some configurations, be loose discs which can rattle in gaps between the others. It is characteristic of the problem that there may be many tight arrangements of discs which have packing values quite close to each other but where the discs cannot be moved from one configuration to another without shrinking them a little. That is, there may be many local minima in configuration space, not always accessible one from another. It is interesting to represent the positions of the points graphically on the screen, with the stereographic projection, to follow how the program moves them about.

The first task in tackling the problem is to find a strategy for doing it, an algorithm. We may try the basic strategy of throwing the points at random on to the surface of the sphere, expanding them up until they touch and then moving them around collectively to let them expand to the maximum without overlapping. This would correspond to some physical experiment, for example with soap bubbles. The difficult task of escaping from one local solution to a globally better solution comes later. Again, various techniques for doing this have been developed, but there is no general rule.

The algorithm is then

(1) Choose N. Put N points at random on to the surface of a unit sphere. Define arrays $\theta(N)$ and $\phi(N)$ for the co-latitude and for the longitude of the N points.

(2) Define a sub-routine to find the angular (great circle) distances between all pairs of points. This uses the cosine rule for spherical triangles.

(3) Define a sub-routine to find those pairs of points which are closer together than the target distance D.

(4) Define a sub-routine to move apart those pairs of points which are closer together than the target distance D so that they become the distance D apart. Each point of a pair is to be shifted a distance $D/2$ in the direction of the line joining the pair. It may be convenient to apply only a fraction of the calculated correction.

(5) For each point, sum vectorially the corrections to be applied to it. The corrections are then applied simultaneously to all points.

(6) The whole process is repeated. The target distance D can be adjusted manually or an automatic procedure can be devised.

(7) If facilities are available it is convenient to draw on the VDU screen a stereogram of the sphere (in upper and lower hemispheres) to show the progress of the refinement. A segment or sub-routine is needed for this.

(8) A sub-routine for recording the θ and ϕ coordinates of each point and the distances between them is necessary for reference.

A program PACK for examining this problem is given in the appendix, but some of the processes are of general interest and will be discussed further.

5.7 To Arrange Points at Random on a Sphere

It is a theorem, known since Archimedes, that if a sphere is sliced into layers of equal thickness, each slice has the same area of curved surface. The obliquity of the surface compensates for the reduction in radius. Accordingly, the following program segment will allocate random spherical coordinates θ, ϕ to N points as a starting configuration.

```
REM θ and φ in radians
DIM T(100), F(100)
N=50
PI=3.1415926
REM define inverse cosine (fails at x=−1)
DEF FNC(X)=2*ATN(SQR((1−X)/(1+X)))
```

```
FOR I=1 TO N
F(I)=2*PI*RND(0)
T(I)=FNC(2*RND(0)-1)
NEXT I
```

The construction of a stereogram (a stereographic projection of the sphere on to a disc in a plane) was described in section 5.3.

6 APL — a higher level language

APL (short for A Programming Language) was developed by K. Iverson from 1962, to be a computer language which would operate with multi-dimensional arrays as single symbols. It is one of the highest level languages generally available and matrix algebra is handled extremely concisely. APL is normally an interpreted language, where lines are executed in succession, and was designed to be used interactively. At first it was implemented on an IBM 360 system, and then on dedicated IBM minicomputers, but now it has reached a stage where it has been provided in a form whereby one can have a full system on a micro-computer at very low cost.

This has brought a huge range of possibilities within the reach of the average private scientific user. It makes the use of matrices and vectors very easy for calculation and not just as a formalism. Besides the mathematical applications discussed here, APL can also be used for logical manipulations and even for business applications.

A complete range of mathematical functions is provided together with a number of functions for manipulating arrays. The most powerful facilities are those for forming the generalised inner and outer products of arrays and for finding the left inverse of a rectangular array. The latter function permits the solution of a system of linear equations with a single simple statement (for example, [A] [X] = [H] is solved (for the least squares solution − [A] need not be square), as X is H matdiv A, that is, $[X] = [H] [A]^+$, where all the variables are arrays and where $[A]^+$ is the generalised inverse of [A]).

In the latest version, APL2 (not yet available for micros), the zeros of a polynomial can be found automatically and the eigenvalues and vectors of a matrix are also found with single operations. Here we give separate programs which can be used to do these operations. Using these programs will lead to an appreciation of the possibilities of APL.

The normal versions of APL use special symbols, many from the Greek alphabet, but in the version available (from MicroAPL Ltd) these over-cryptic characters are represented by words which have more mnemonic value. The normal user will find these much more convenient, since one of the difficulties is that APL can be, although need not be, written extremely densely and it may be far harder than for any other language to see what is happening. An example is given in the next section which shows the conciseness of APL in comparison with BASIC.

6.1 To Fit a Polynomial to a Curve

Having acquired programs for calculating the inverse or the generalised inverse
of a matrix, these can be used for a great variety of purposes where linear
equations may be solved by least squares.

For example, we may have a table of values of Y as a function of X and wish
to find the best polynomial – straight line, parabola or cubic (as polynomials
of order 1, 2, 3 respectively) – which will fit these observed points with a
smooth curve. We have a table $Y_{obs.}$ of observed values of Y as a function of X
and, for the same values of X, wish to calculate values of Y, $Y_{calc.}$, so that

$$\Sigma(Y_{obs.} - Y_{calc.})^2$$

is a minimum. We may attach weights to the equations according to our confi-
dence levels.

$$Y_{obs.} = a_1 X^0 + a_2 X^1 + a_3 X^2 + a_4 X^3 \ldots$$

is thus one of a series of linear equations, one for each observed point, for the
coefficients $a_1 \ldots a_4$ to be solved by least squares. Here we are solving for
a_1, a_2, a_3, a_4, the coefficients of successive powers of X (and not for X, the
character used normally to represent an unknown).

In APL the answer can be obtained in one line

A is Y matdiv X outer.exp 0 1 2 3

(In APL a series of numbers are separated by spaces instead of commas.) This
means, raise each value of X in turn to the powers 0, 1, 2, 3, giving an $n \times 4$
matrix. *Outer* means take each value of X with each of the exponents 0, 1, 2, 3.
Matdiv X means form the left inverse (a $4 \times n$ matrix) of the $n \times 4$ matrix just
obtained. This is then multiplied by the n-dimensioned vector of the Y values
to give the four values of a_i.

More fully, using BASIC, we may set up the array M

$$
\begin{matrix}
X_1^0 & X_1^1 & X_1^2 & X_1^3 \\
X_2^0 & X_2^1 & X_2^2 & X_2^3 \\
\cdot & \cdot & \cdot & \cdot \\
\cdot & \cdot & \cdot & \cdot \\
\cdot & \cdot & \cdot & \cdot \\
X_n^0 & X_n^1 & X_n^2 & X_n^3
\end{matrix}
$$

for the n observed points, find its generalised inverse M^+, and then obtain the
value of A by matrix multiplication as $A_i = M^+_{ij} Y_j$ or, conventionally and faster,
we may use the ordinary inverse to invert the square matrix $[M^T M]$ obtaining
the coefficients A by $A = [M^T M]^{-1} A^T Y$.

6.2 To Fit a Polynomial to a Surface

The previous procedure can be extended to two or more dimensions to fit, for example, a quartic surface (an ellipsoid or hyperboloid) to an array of points – 'spot heights' – Z given as a function of X and Y

$$Z_{obs.} = F(X, Y)$$

The equation of the quartic contains the first and second powers of X and Y with six necessary coefficients. $Z_{calc.}$ may be found as a quartic by

$$Z_{calc.} = a_1 + a_2 X + a_3 Y + a_4 X^2 + a_5 XY + a_6 Y^2$$

and these coefficients $a_1 \ldots a_n$ can be found by least squares as before, provided we have at least six points.

6.3 Curve Fitting

We may distinguish several cases: (a) to fit a straight line to points by least squares; (b) to fit a polynomial to points by least squares; (c) to fit some other function to points by least squares; (d) to fit exactly by a polynomial of order n (through $n + 1$ points); and (e) to fit exactly by segments of a cubic.

6.4 Splines

A spline is a way of fitting a number of curved segments through designated points so that there is no discontinuity of gradient or curvature (first or second derivatives). This requires a series of cubic segments. We set up a number of linear equations. We have the y-coordinates of the i points at the ordinate x

$$y_i = A_i + B_i x + C_i x^2 + D_i x^3$$

then the gradients at these points – the first derivatives

$$y_i' = \quad B_i + 2C_i x + 3D_i x^2$$

and finally the curvatures, the second derivatives at these points

$$y_i'' = \quad\quad\quad 2C_i + 6D_i x$$

For four points we have (figure 6.1) positions

$$y_1 = A_{12} + B_{12} x_1 + C_{12} x_1^2 + D_{12} x_1^3$$
$$y_2 = A_{12} + B_{12} x_2 + C_{12} x_2^2 + D_{12} x_2^3$$
$$y_2 = A_{23} + B_{23} x_2 + C_{23} x_2^2 + D_{23} x_2^3$$
$$y_3 = A_{23} + B_{23} x_3 + C_{23} x_3^2 + D_{23} x_3^3$$

$$y_3 = A_{34} + B_{34}x_3 + C_{34}x_3^2 + D_{34}x_3^3$$

$$y_4 = A_{34} + B_{34}x_4 + C_{34}x_4^2 + D_{34}x_4^3$$

Gradient at 2: $B_{12} + 2C_{12}x_2 + 3D_{12}x_2^2$

$$= B_{23} + 2C_{23}x_2 + 3D_{23}x_2^2$$

Gradient at 3: $B_{23} + 2C_{23}x_3 + 3D_{23}x_3^2$

$$= B_{34} + 2C_{34}x_3 + 3D_{34}x_3^2$$

Curvature at 2: $2C_{12} + 6D_{12}x_2$

$$= 2C_{23} + 6D_{23}x_2$$

Curvature at 3: $2C_{23} + 6D_{23}x_3$

$$= 2C_{34} + 6D_{34}x_3$$

We determine also the gradient at the beginning and the end, that is, $y_i' = m_1$ and $y_4' = m_2$

$$m_1 = B_{12} + 2C_{12}x_1 + 3D_{12}x_1^2$$

$$m_2 = B_{34} + 2C_{34}x_4 + 3D_{34}x_4^2$$

and we have 12 equations for 12 unknowns ($A_{12} \ldots D_{34}$), which can be set up and solved by the standard methods which we have explained.

[To fit a spline though the points $(x, y) = (0, 0), (10, 8), (17, 10), (30, 20)$ with an initial gradient of 2 and a final gradient of 4. We find $A_{12} = 0, B_{12} = 2$, $C_{12} = -0.197552, D_{12} = 0.0077552; A_{23} = 14.6029, B_{23} = -2.38086$, $C_{23} = 0.240533, D_{23} = -0.00684765; A_{34} = -86.54, B_{34} = 15.4679$, $C_{34} = -0.0809392, D_{34} = 0.0137391$.]

[To fit a polynomial of degree 3 through the same points with these gradients we add two points to define the gradients making: $(0, 0), (0.5, 1), (10, 8)$, $(17, 10), (20.5, 18), (30, 20)$ and find the coefficients, $0, 1.41095, -0.0792372$, 0.00181426.]

6.5 To Fit a Polynomial

This is done particularly expeditiously using APL. For example, if we wish to fit a polynomial of degree 5 through six points (figure 6.2) then we have

    Y is 0 1 8 10 18 20   *X* is 0 0.5 10 17 29.5 30

    A is Y matdiv X outer.exp 0 1 2 3 4 5

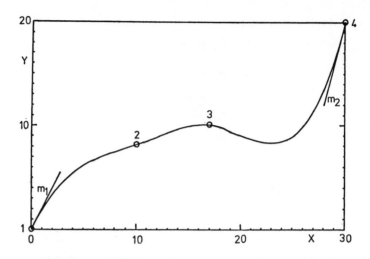

Figure 6.1. An example of a spline. Three cubic segments, 12, 23 and 34 are fitted by least squares through the points 1, 2, 3, 4 (0,0; 10,8; 17,10; 30,20) so that the gradients and curvatures match at each point. Initial and final gradients $m_1 = 2$ and $m_2 = 4$.

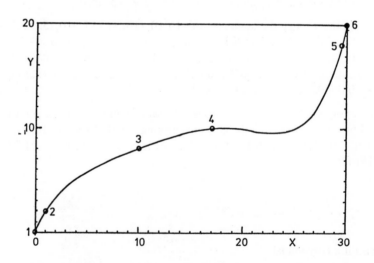

Figure 6.2. A fifth-degree polynomial (6 parameters) is fitted by least squares through the points 1, 2, 3, 4, 5, 6 as in figure 6.1. Points 2 and 5 are added to define the required gradients. In APL, given $y = a_1 x^0 + a_3 x^2 + \ldots + a_6 x^5$, the statement is: *A is Y MATDIV X outer.exp 0 1 2 3 4 5. Y_{calc}.* is found from: *YCALC is (X outer.exp 1 2 3 4 5) +. *A.*

which gives the coefficients A as

   0, 2.14617, −0.305216, 0.0263073, −0.00109082, 1.6608 E − 5.

This is checked by: *ycalc is (X outer.exp 0 1 2 3 4 5) + . *A* which returns
0 1 8 10 18 20.

7 The partition of space into domains

For ecological, geographical, administrative, chemical and many other purposes, it may be necessary to divide space (of 2 or 3 or indeed of N dimensions) containing a number of points, such as the centres of atoms, the nests of birds, market towns, sources of water etc., so that the whole of space is allocated to the domain of one point or another and no space is left unallocated. The word used for domain in the German literature (and sometimes taken over untranslated) is *Wirkungsbereich* − region of influence.

7.1 Voronoi Polygons (or Polyhedra)

The simplest example is the division of a plane containing centres into Voronoi polygons, one surrounding each point, so that every point in a particular (necessarily convex) polygon is nearer to its centre than to any other centre. This dissection, first described in 1908, is called after G. F. Voronoi. The method of achieving this is to draw the perpendicular bisector between every pair of centres and to take as the Voronoi polygon of a particular centre, the inmost segments of these bisecting lines. In this way the whole plane can be divided into convex polygons each containing one centre (figure 7.1).

There is a variety of algorithms for carrying through this process automatically.

The basis consists of identifying possible vertices for the polygon which are the circumcentres of triangles defined by the three points at their vertices. There must be no other point nearer to the circumcentre than the three points which define the triangle. The network of triangles is called the Delone tessellation.

In fact any unique point in a triangle, besides the circumcentre mentioned above, will give a type of dissection with linear interfaces. The most important alternative is the radical line dissection, convenient when each centre is not just a point, but has a characteristic radius or range of action (as an atom or a city or a plant has). The radical line defined by two such circles is the locus of all points from which the tangents to the two circles are of equal length. It is a straight line and when the two circles are touching it becomes the common tangent. If the circles overlap, the tangent length becomes imaginary in some places but the radical line becomes the common chord. In our example of the radical plane dissection the function

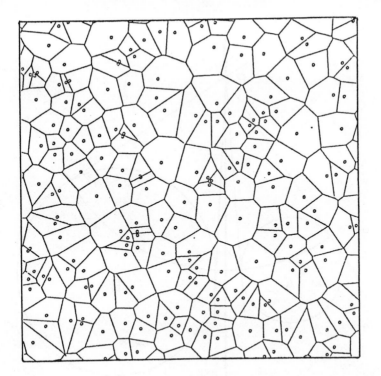

Figure 7.1. Illustration of Voronoi dissection (after Hasegawa and Tanemura (1980)).

$$T^2 = (x(i) - xp)^2 + (y(i) - yp)^2 - R(i)*R(i)$$

is used and the centre with the least value of T^2 (which may be negative) is taken as the nearest (figure 7.2).

7.2 Methods of Calculation

In general there are two main approaches, exact and statistical. The exact methods consist in computing the vertices of the polygons and keeping careful account of the status of each vertex (which points took part in defining it). The statistical method consists in taking points at random in space and finding which centres are nearest to each, and gradually building up the picture.

As the number of pixels in the display being used decreases, the statistical methods become increasingly competitive with the exact methods and, for microcomputers, are usually to be recommended for cost-effectiveness where the writing time of programs is large compared with the expected time of

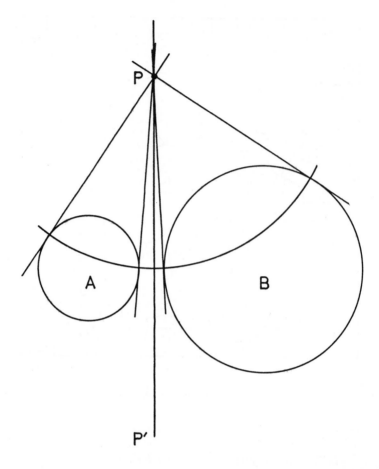

Figure 7.2. The radical plane between two spheres is the locus of all point from which the lengths of tangents to the two spheres are equal.

usage. Indeed, for the lowest resolution displays (such as the BBC computer's MODE 2 which has 160 × 256 elements and 16 colours) the brute force algorithms where the calculation of which point is nearest is done for each of the pixels. It is entirely a matter of estimating the relative costs of production time, equipment time, writing and development time.

7.3 Statistical Methods

Statistical methods are by far the easiest and the most flexible to apply and it is often convenient to begin to examine a problem in this way. If necessary, more exact methods can later be deployed. A program VORON in the appendix gives an example.

It is necessary first to define a 'generalised potential', the effect due to a centre of weight W at a distance D. Then we examine in turn the potentials due to each centre at a randomly chosen point. The point is then allocated (for example, by marking it on the screen with an appropriate colour) to the domain of a particular centre. All kinds of criteria of choice are clearly possible. For such general laws of allocation the frontiers between domains need not be planar, the centres of power need not lie within their domains and the domains themselves need not be connected (and may be in several separated parts). Statistical methods are then very suitable for exploring the effects of different 'generalised potentials' on the shapes of domains. Clearly, in for example geographical problems, the criteria may be modified by boundary conditions in an arbitrary way. For example, in geography, 'distance to market' may be strongly modified by obstacles.

If we consider the approach of dividing space by planes then the perpendicularly bisecting plane and the radical plane dissections give unique answers, but other planes defined differently may leave regions unallocated. The generalised potential approach mentioned above is the most satisfactory.

Two simple techniques are possible. (a) To choose test points at random over the area. If the sampling of points is less than about 25 per cent of the total then the technique is very effective but clearly if all point are eventually to be examined the efficiency drops off. Even a few points (1000) rapidly begin to outline the character of the dissection. (b) To compute the generalised potentials at all points.

7.4 Exact Calculation

The Voronoi dissection of space (in three dimensions) has the following properties:

(a) All space is allocated so that the total volume is the sum of the volumes of the individual domains.
(b) Every point in a domain is nearer to the centre defining the domain than to any other centre.
(c) The partitions between the domains are planar (and are the planes perpendicularly bisecting the lines joining pairs of centres).
(d) The domains are convex polyhedra.
(e) The partitioning is unique.
(f) Centres at the surface of a finite cluster may have domains which are unbounded or abnormally large. That is, the procedure for dealing with the surface of a cluster is undefined.

(g) The dissection can be extended for use as a diagram of forces (Maxwell, 1869).

(h) Each vertex of the Voronoi polyhedron is the circumcentre of a tetrahedron of four defining centres. No other centre is nearer than these four, although, if there is exact symmetry, other centres may also lie on the same circumsphere.

(i) This degeneracy may be removed by slight perturbations of position or full provision may be made in the program.

(j) If we take all the tetrahedra, the centres of which are valid Voronoi vertices, then this is the associated Delone tessellation (called after B. N. Delone).

(k) It is unique, unless there is symmetrical degeneracy as described above, in which case several choices of tetrahedra are equivalent. It also fills all space.

(l) A unique definition of coordination follows from the Voronoi dissection: given a centre and its Voronoi polyhedron, any other centre, participating in the definition of a vertex of this polyhedron, is deemed to be coordinated.

(m) The sum of the volumes of all the Delone tetrahedra about each centre is equal to four times the sum of the volumes of the Voronoi polyhedra (since each tetrahedron is counted for the centre at each of its four vertices).

The Voronoi dissection is of reduced utility when dealing with centres, such as atoms, which may be of different known sizes. In such a case the radical plane (the locus of equal tangent lengths) may be used instead of the perpendicularly bisecting plane. The exact construction of a Voronoi dissection requires careful accountancy. J. L. Finney (1979) has given a clear account of the procedure and Brostow *et al.* (1978) give a procedure which, since it takes a computing time proportional to $N^2 \log N$, instead of N^4 as for Finney's program, should be faster for large assemblies. Finney's program nevertheless competes by reducing the volumes over which possible coordinating atoms are sought using a knowledge of the reasonableness of the structure of real materials.

8 The best fit between two shapes or molecules

It is often necessary to compare two objects, such as a pair of molecules, where one array of numbered points in space is to be matched against another *correspondingly numbered* set. There are several other more difficult cases where corresponding points are not numbered or perhaps even not identified as points but are only part of a continuous density distribution. A considerable number of methods have been developed.

Suppose that we have two sets of corresponding coordinates of N points in space, $A<N,3>$ and $A'<N,3>$. The centres of gravity of both have first been reduced to a common origin. One definition of centre of gravity is that point about which the sum of the squares of the distances of the atoms from the point is a minimum. We wish then to find the 'best' rotation matrix relating the two molecules by $A<N,3>R<3,3>=A'<N,3>$.

If $R<3,3>$ represents a pure rotation then it contains only three independent quantities (two quantities to specify an axis and one for the amount of rotation about this axis). These are difficult to handle symmetrically and it may be satisfactory simply to find the 'best' matrix $R<3,3>$ with nine quantities which relates the two molecules. Suppose that we solve the equations $A<N,3>R<3,3>$ $=A'<N,3>$ for R by least squares (minimising the sum of the squares of the discrepancies of the N equations). The solution is

$$R<3,3>=[A^T<3,N>A<N,3>]^{-1}A^T<3,N>A'<N,3>$$

Apart from the matrix multiplication we only have to invert a 3 x 3 matrix, which can be done explicitly.

Having found $R<3,3>$ (with nine components) we may ask what it means. We may consider this in two different ways: we may divide $R<3,3>$ into two parts, one symmetrical RS and the other anti-symmetrical RA, and consider them separately. The anti-symmetrical part represents a pure rotation and the symmetrical part represents a deformation of a sphere into an ellipsoid.

$$R<3,3>= RA<3,3>+ RS<3,3>$$
$$RS<3,3>=[R<3,3>+R^T<3,3>]/2$$
$$RA<3,3>=[R<3,3>-R^T<3,3>]/2$$

8.1 Quaternions

Here we venture to introduce a topic, considered as advanced, namely that of
quaternion algebra. We do this because we believe that this subject, popular in
the Victorian period, will soon revive because it provides a superior method of
handling the rotations of solid objects and will thus take a place in the repertoire
of the programmer.

We have seen (section 2.3) that the algebra of complex numbers, a pair of
numbers manipulated together, is very suitable for plane geometry, where a
point requires two coordinates. Of a complex number, $z = x + iy$, x is the real
part and y the imaginary, measured along the real axis in real numbers and
along the imaginary axis at right angles to it in units of i respectively. i has the
property that $i^2 = -1$ but nothing else. Any expression using complex numbers
can be separated by appropriate multiplication into its real and imaginary parts.
We may also have the complex conjugate of z, denoted by z^* which is $x - iy$,
so that zz^* is $x^2 + y^2$, the square of the amplitude or length of z. Complex
numbers can be divided and, by multiplying numerator and divisor appropriately
by the same quantity, usually the complex conjugate of the denominator, can
be separated into real and imaginary parts. A rotation of $90°$ can be achieved
by multiplication by i, and rotation of θ by multiplying by the complex number
$\cos \theta + i \sin \theta$. Vectors cannot be divided.

On 16 October 1863, Sir William Rowan Hamilton was standing on the
Brougham Bridge over the Royal Canal in Dublin, pondering on how to extend
the algebra of complex numbers to three dimensions, when the idea of quaternions struck him. He engraved the rule on a stone on the balustrade. It was
$i^2 = j^2 = k^2 = ijk = -1$.

Hamilton's quest was for an algebra which would do the same kind of thing
for three dimensions. Obviously this was much more complex but eventually
he found the solution. This consisted of an algebra of sets of four numbers,
quaternions, reviving a venerable English word. The rules of quaternion algebra
go as follows.

A quaternion consists of four numbers (q_s, q_x, q_y, q_z), which are manipulated together. The first, q_s, is a scalar and measures units, dimensionless
numbers. The other three are the components of a vector $\mathbf{q}$. These three vector
components measure units along the three perpendicular coordinate axes. Unit
vectors along these axes are $\mathbf{i}, \mathbf{j}, \mathbf{k}$, respectively. Quaternions may be added
and subtracted simply by adding or subtracting corresponding components,
but the rules for multiplication are more complicated. Multiplication is non-
commutative, that is, PQ is not equal to QP. This corresponds to the algebra
of rotations, because a rotation of $90°$ clockwise (looking in towards the origin)
about the X-axis followed by a rotation of $90°$ clockwise about the Y-axis does
not produce the same result as the same two operations performed in the reverse
order. Try it by turning a box on the table.

A body may be described in terms of vectors from the origin to each identified point in it. In rotating the body we wish to change the old vectors to new

vectors which correspond to the rotated orientation. Quaternion algebra permits us to update these position vectors. A quaternion can most conveniently be used to represent a rotation through an angle θ about a specified direction. Let us concentrate on describing quaternions in terms of this example, which is the most important application. The vector r is to be rotated through an angle θ (clockwise when looking towards the origin along the rotation axis) about the axis with direction cosines l, m, n (where $l^2 + m^2 + n^2 = 1$) to become a vector r'. This rotation may need to be followed by several more in concatenation (in a chain). There are various ways in which this may be done but that recommended by Rooney (1977) is to split the operation and to multiply twice as in the equation $r = Q^{-1}rQ$. Here Q^{-1} is the quaternion inverse to Q and Q is the quaternion $\cos(\theta/2) + \sin(\theta/2)n$, where n is the unit vector with direction cosines l, m, n. $Q^{-1} = \cos(\theta/2) - \sin(\theta/2)n$. When this expression is multiplied out we have

$$r' = (\cos^2(\theta/2) - \sin^2(\theta/2))r - (n \times r)\sin\theta + (n.r)n\sin^2(\theta/2)$$

where the scalar part is zero, as it should be for a vector result. This expression combines the scalar and vector products of n and r.

The rotation operation is entirely equivalent to that achieved by multiplying the vector r by the complicated matrix R

$$[R] = \begin{bmatrix} ll(1-\cos\theta)+\cos\theta, & ml(1-\cos\theta)+n\sin\theta, & nl(1-\cos\theta)-m\sin\theta \\ lm(1-\cos\theta)-n\sin\theta, & mm(1-\cos\theta)+\cos\theta, & nm(1-\cos\theta+l\sin\theta \\ ln(1-\cos\theta)+m\sin\theta, & mn(1-\cos\theta)-l\sin\theta, & nn(1-\cos\theta)+\cos\theta \end{bmatrix}$$

where l, m, n are the direction cosines of the axis of rotation, and θ is the angle of rotation about that line.

Two quaternions give a (non-commutative) product which is a quaternion. Hamilton's conditions, $i^2 = j^2 = k^2 = -1$, $ij = -ji$, $ij = k$, etc. are applied to simplify the product. The quaternion product of two vectors R_1 and R_2 is

$$R_1 R_2 = -R_1.R_2 + R_1 \times R_2$$

where the terms are the normal scalar and vector products. The product of two quaternions P and Q is

$$PQ = (p_s q_s - p.q) + (p_s q + q_s p + p \times q)$$

The product of a quaternion with a vector is

$$Qr = (-q.r) + (q_s r + q \times r) \quad \text{and}$$

$$rP = (-r.p) + (p_s r + r \times p) \quad \text{so that}$$

$$QrP = [-q_s(p.r) - p_s(q.r) - q.r \times p] \quad \text{(the scalar part)}$$
$$+ [q_s p_s r + q_s(r \times p) - r.p)q$$
$$+ p_s(q \times r) + (q.p)r - (q.r)p] \quad \text{(the vector part)}$$

Thus, if $Q = \cos\theta/2 + \sin\theta/2\,\boldsymbol{n}$ and $Q^{-1} = \cos\theta/2 - \sin\theta/2\,\boldsymbol{n}$

$$\boldsymbol{r}'=Q^{-1}\boldsymbol{r}Q = (\cos^2\theta/2 - \sin^2\theta/2)\,\boldsymbol{r} - \sin\theta\,(\boldsymbol{n}\times\boldsymbol{r}) + 2\sin^2\theta/2\,(\boldsymbol{n}.\boldsymbol{r})\boldsymbol{n}$$

the scalar part is zero. This expression is needed for calculating $\boldsymbol{r}'$ from $\boldsymbol{r}$ for each position vector when θ and l, m, n have been found.

To find θ and l, m, n we begin with $Q\,\boldsymbol{r}' = \boldsymbol{r}Q$ (since $Q^{-1}Q = 1$) so that, equating parts

$$\boldsymbol{q}.\,(\boldsymbol{r}'-\boldsymbol{r}) = 0$$

$$q_s = \cos\theta/2 \text{ and } \boldsymbol{q} = \sin\theta/2$$

so that

$$(\boldsymbol{r}-\boldsymbol{r}') = \tan\theta/2\,\boldsymbol{n}\times(\boldsymbol{r}+\boldsymbol{r}')$$

Converting to orthonormal coordinates

$$(x-x')\boldsymbol{i} + (y-y')\boldsymbol{j} + (z-z')\boldsymbol{k} = \tan\theta/2 \begin{bmatrix} \boldsymbol{i} & \boldsymbol{j} & \boldsymbol{k} \\ l & m & n \\ (x+x') & (y+y') & (z+z') \end{bmatrix}$$

So that, finally, we have three linear equations for each pair of related atoms and thus $3N$ in all (if there are N pairs). Writing t for $\tan\theta/2$

$$mt(z+z') - nt(y+y') = (x'-x)$$

$$-lt(z+z') \qquad\quad + nt(x+x') = (y'-y)$$

$$lt(y+y') - mt(x+x') \qquad\quad = (z'-z)$$

These $3N$ equations are then solved by the least squares procedure, $[A][X]=[H]$ giving $[X] = [A^TA][H]$ for the three unknowns, lt, mt and nt. $\tan^2\theta/2$ is obtained by squaring and adding these solutions. Division by t then gives l, m, n, the direction cosines. A value of θ between $0°$ and $180°$ is obtained, the procedure failing if $\theta = 180°$, when $\tan\theta/2$ is infinite. When this case is detected a special procedure must be implemented as $Q = (1, 0, 0.\ 0)$ which is the same as the identity. Since the case of $180°$ rotation is frequent, this is the main disadvantage of the method. However, this defect is readily obviated, if it causes trouble, by initially rotating the stationary molecule by an arbitrary amount, for example, by $30°$, before setting up the equations, and then after solution, rotating it back again. It may also be necessary to change the sign of θ since the positive square root is returned.

A straightforward program (QUAT) has been written to follow the procedure described. It is listed in the appendix. The two molecules or shapes are referred to their centres of gravity and the vector to each atom from the c.g. are normalised to unity, the lengths being retained as weights for the least squares procedure. Clearly other weighting schemes could be applied also. The $3N$ linear equations are then set up and solved by least squares, the inversion of the 3×3 matrix being written explicitly. Having obtained θ and l, m, n the second molecule or shape can be rotated to the orientation of the first for comparison, the difference in the positions of corresponding atoms being printed out.

Relevant references are Rooney (1977) and Mackay (1984).

8.2 The Transformation of Non-orthogonal Axes (in N-dimensions) to Orthogonal Axes. The Gram–Schmidt Procedure

If we are working with the natural symmetry axes of a crystal structure, which may be oblique, or have been using the procedure for finding Cartesian coordinates when given the set of distances between members of a set of points, it will be necessary to transform from the oblique axes to the usual, orthogonal Cartesian axes. This is done with the Gram–Schmidt procedure.

Given a non-orthogonal set of base vectors a_i, these axes can be transformed to the mutually orthogonal set b_i by the sequence of operations

$$b = a_i - \Sigma((b_j.a_i)/(b_j.b_j)) \, b_j$$

The new axial lengths are normalised to the orthonormal set of axes c_i by

$$c_i = b(b.b)^{\frac{1}{2}}$$

Taking the example of triclinic axes a, b, c, we have

$$A = a$$

$$B = b - (A.b/A.A) \ A$$

$$= b - (a.b/a.a) \ a$$

(That is, in the plane of b and a but perpendicular to a. A and B are thus perpendicular since $A.B = a.b - (a.b/a.a)a.a = 0$.)

$$C = C - (A.c/A.A) \ A - (B.c/B.B) \ B$$

Gram's determinant is the determinant of the metric matrix, the array of inner products of the axes, in three dimensions

$$\begin{bmatrix} a.a & a.b & a.c \\ b.a & b.b & b.c \\ c.a & c.b & c.c \end{bmatrix}$$

This is the square of the triple product $[a \ b \ c]$ and thus equal to the square of the volume of the parallelepiped or unit cell.

The basic problem of distance geometry is to transfer from a set of distances between points to Cartesian coordinates. If we have a number of points (in space of N dimensions) then we may attach orthogonal coordinates to them so that

(1) is at $0,0,0,0,0,0$
(2) is at $x_2,0,0,0,0,0$
(3) is at $x_3,y_3,0,0,0,0$
(4) is at $x_4,y_4,z_4,0,0,0$
(5) is at $x_5,y_5,z_5,0,0,0\ldots$
etc.

$$
\begin{bmatrix} 0 & 0 & 0 & 0 \\ x_2 & 0 & 0 & 0 \\ x_3 & y_3 & 0 & 0 \\ x_4 & y_4 & z_4 & 0 \end{bmatrix} \times \begin{bmatrix} 0 & x_2 & x_3 & x_4 \\ 0 & 0 & y_3 & y_4 \\ 0 & 0 & 0 & z_4 \\ 0 & 0 & 0 & 0 \end{bmatrix}
$$

$$
= \begin{bmatrix} 0 & 0 & 0 & 0 \\ 0 & x_2^2 & x_2 x_3 & x_2 x_4 \\ 0 & x_2 x_3 & x_3^2+y_3^2 & x_3 x_4 + y_3 y_4 \\ 0 & x_2 x_4 & x_3 x_4 + y_3 y_4 & x_4^2 + y_4^2 + z_4^2 \end{bmatrix}
$$

$$x_2^2 = d_{12}^2, \quad x_3^2 + y_3^2 = d_{13}^2, \quad x_4^2 + y_4^2 + z_4^2 = d_{14}^2.$$

$$x_2 x_3 = d_{12}.d_{13}$$

Start with a matrix of the distances d_{ij}^2. Take point 1 as the origin. Subtract the first row from each row and the first column from each column. The terms are then

$$
\begin{bmatrix} 0 & d_{12}^2 & d_{13}^2 & d_{14}^2 \\ d_{21}^2 & 0 & d_{23}^2 & d_{24}^2 \\ d_{31}^2 & d_{32}^2 & 0 & d_{34}^2 \\ d_{41}^2 & d_{42}^2 & d_{43}^2 & 0 \end{bmatrix}
$$

The modified distance matrix can then be factorised into upper and lower triangular matrices with zeros on the diagonal. A symmetrical positive definite matrix can be factorised into the product of a lower triangular matrix and its transpose by the Choleski method. To be 'positive definite' means that every symmetrical determinant which can be constructed from the original matrix by symmetrically deleting rows and columns is $\geqslant 0$.

If we require a preliminary solution it may be allowable to change the signs of certain terms and deal with a matrix which is not quite positive definite. Many physical situations give rise to matrices which are essentially positive definite, for example the distances between points in 2-dimensional or 3-dimensional space. Another technique is to put large terms along the diagonal of the matrix. This has been done in the program CHOLESKI (see appendix).

The matrix is factorised in order as follows

$$
\begin{bmatrix}
0 & a_{12}^2 & a_{13}^2 & a_{14}^2 \\
a_{21}^2 & 0 & a_{23}^2 & a_{24}^2 \\
a_{31}^2 & a_{32}^2 & 0 & a_{34}^2 \\
a_{41}^2 & a_{42}^2 & a_{43}^2 & 0
\end{bmatrix}
$$

$$
=
\begin{bmatrix}
L_{11} & 0 & 0 & 0 \\
L_{21} & L_{22} & 0 & 0 \\
L_{31} & L_{32} & L_{33} & 0 \\
L_{41} & L_{42} & L_{43} & L_{44}
\end{bmatrix}
\times
\begin{bmatrix}
L_{11} & L_{21} & L_{31} & L_{41} \\
0 & L_{22} & L_{32} & L_{42} \\
0 & 0 & L_{33} & L_{43} \\
0 & 0 & 0 & L_{44}
\end{bmatrix}
$$

$$a_{11} = L_{11}^2 \qquad\qquad\qquad L_{11} = (a_{11})^{\frac{1}{2}}$$

$$a_{21} = L_{21}L_{11} \qquad\qquad\qquad L_{21} = a_{21}/L_{11}$$

$$a_{22} = L_{21}^2 + L_{22}^2 \qquad\qquad L_{22} = (a_{22} - L_{21}^2)^{\frac{1}{2}}$$

$$a_{31} = L_{31}L_{11} \qquad\qquad\qquad L_{31} = a_{31}/L_{11}$$

$$a_{32} = L_{31}L_{21} + L_{32}L_{22} \qquad L_{32} = (a_{32} - L_{31}L_{21})/L_{22}$$

$$a_{33} = L_{31}^2 + L_{32}^2 + L_{33}^2 \qquad L_{33} = (a_{33} - L_{31}^2 - L_{32}^2)$$

$$a_{41} = L_{41}L_{11} \qquad\qquad\qquad L_{41} = a_{41}/L_{11}$$

$$a_{42} = L_{41}L_{21} + L_{42}L_{22} \qquad L_{42} = (a_{42} - L_{41}L_{21})/L_{22}$$

$$a_{43} = L_{41}L_{31} + L_{42}L_{32} + L_{43}L_{33} \qquad L_{43} = (a_{43} - L_{41}L_{31} - L_{42}L_{32})/L_{33}$$

$$a_{44} = L_{41}^2 + L_{42}^2 + L_{43}^2 + L_{44}^2 \qquad L_{44} = (a_{44} - L_{41}^2 - L_{42}^2 - L_{43}^2)^{\frac{1}{2}}$$

We follow the steps down the right-hand column. Since this involves division by the diagonal term a_{11} this term should not be zero. For example

$$
\begin{bmatrix}
1 & 2 & 5 \\
2 & 13 & 31 \\
5 & 31 & 195
\end{bmatrix}
=
\begin{bmatrix}
1 & 0 & 0 \\
2 & 3 & 0 \\
5 & 7 & 11
\end{bmatrix}
\times
\begin{bmatrix}
1 & 2 & 5 \\
0 & 3 & 7 \\
0 & 0 & 11
\end{bmatrix}
$$

As an example of this procedure we took the data provided by the air fares in dollars between 10 European cities (in the period before the single fare structure had broken up!) as rough measures of the distances between them. The problem is to reconstruct the map of Europe from this rough data. For 10 cities we have $(10 \times 2) - 3 = 17$ unknowns to be determined by $10 \times 9 / 2 = 45$

observations. We have the guarantee that there is in fact a real physical map as the answer.

The data was

	BRU	LON	PAR	FRA	MAD	MOS	ROM	VIE	WAR	ZUR
BRUSSELS	—	35.2	30.3	26.6	75.6	161.2	70.6	68.3	83.9	38.1
LONDON		—	31.0	53.9	87.3	185.4	98.8	93.9	119.2	64.6
PARIS			—	39.8	67.0	167.8	68.6	72.6	99.0	39.4
FRANKFURT				—	81.0	152.3	63.0	47.8	73.7	29.3
MADRID					—	219.4	70.2	110.2	146.4	76.0
MOSCOW						—	164.8	120.0	91.7	152.3
ROME							—	64.8	106.2	53.1
VIENNA								—	52.8	45.5
WARSAW									—	86.2
ZURICH										—

A Choleski factorisation of this distance matrix was carried out. The initial matrix is not necessarily positive definite so that in order to get a set of starting positions it was necessary arbitrarily to change the signs of negative terms so that a square root could be taken. A set of positions was then obtained

```
 0,   0,       0,   0,   0,   0,   0,   0,   0
35.2,  0,       0,   0,   0,   0,   0,   0,   0
16.9, 25.1,    0,   0,   0,   0,   0,   0,   0
-13.6 10.1     20.5,  0,   0,   0,   0,   0,   0
-9.4 49.1     -33.7  45.5,  0,   0,   0,   0,   0
-101.5 43.7    -3.6 -251.6 222.6,  0,   0,   0,   0
-50.2 57.8    -19.7 -24.1 -53.3 68.6,  0,   0,   0
-41.3 34.2     30.9 -42.0 -36.0 -53.4 71.4,  0,   0
-84.2 20.3     -9.4 1.2 14.1 -64.5 -32.4  77.3,  0
-21.0 30.5     2.75 -211.3 -244.9 -278.8 -456.3 -399.4 741.8
```

The first two columns then give the starting x, y coordinates which were used for refinement to fit the distances. The first three places, BRU, LON and PAR, were fixed and their mutual distances were not refined but taken as the datum. The resulting map is shown in figure 8.1.

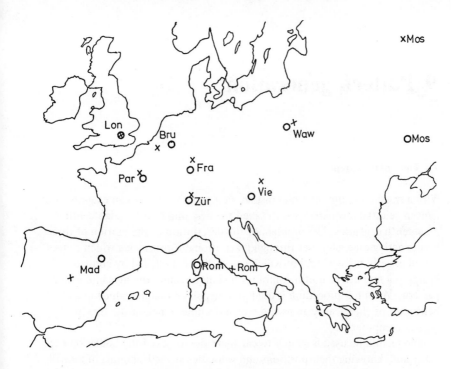

Figure 8.1. Starting with a matrix of the approximate distances between 10 European capitals (as represented by the air fares), orthogonal axes can be found by the Gram-Schmidt method and coordinates with respect to these axes are then refined to fit the distances. The map shows the positions found by this method (circles), compared with the correct positions (crosses).

9 Pattern generation

9.1 The Solar System

With a small computer graphics display or a pen plotter, we can examine problems in celestial mechanics which taxed the best minds of the eighteenth and nineteenth centuries. The problem of the description of the motion of three bodies acting on each other through gravitational forces is one which cannot be solved exactly by algebra, but which can easily be modelled by computer. It brings in a number of useful computer techniques and generates interesting patterns. A similar procedure is used in molecular dynamics which models the behaviour of chemical systems and which occupies much of the time of the biggest computers.

The technique used is simply to calculate the forces of the bodies on each other and, knowing their positions and velocities at the beginning of a small time interval dt, to calculate the new positions and velocities at the end of this time interval. If we mark the position of each body on the screen with a dot, then we have three trajectories which we can observe, the distances between successive points indicating the velocities. If one of the bodies is a spacecraft with rocket propulsion, we can change its velocity by firing the rocket and can see the effects, simulating, for example, the problem of leaving the Earth and landing on the Moon. We take only the case when all bodies are moving in the plane of the display screen.

This procedure is essentially irreversible because the errors accumulate. Laplace's megacomputer project of 1812 would, for this reason, not work. He had suggested: "Given for one instant an intelligence which could comprehend all the forces by which nature is animated and the respective positions of the beings which compose it, if moreover, this intelligence were vast enough to submit these data to analysis, it would embrace in the same formula both the movements of the largest bodies in the universe and those of the lightest atom; to it nothing would be uncertain, and the future as the past would be present to its eyes."

The essential mathematical tool used for finding the change in position is Taylor's Theorem. This says that the value of x at a time $(t + dt)$ is given in terms of the position at time t by: $x(t + dt) = x(t) + dt\, x'(t) + (1/2)(dt)^2 x''(t) + (1/6)(dt)^3 x'''(t) + \ldots$ where $x'(t)$ is the velocity at time t, $x''(t)$ is the acceleration at time t, $x'''(t)$ is the rate of change of acceleration at time t, and so on. If we write down the same expression for the previous time interval $(t - dt)$ to t, and add it to the previous expression, then a number of terms cancel out giving: $x(t + dt) = 2x(t) - x(t - dt) + (dt)^2 x''(t)$. Thus, the new position $x(t + dt)$ equals the old position $x(t)$ plus the change $(x(t) - x(t - dt))$ which took place in the previous interval, plus the acceleration times $(dt)^2$. Since the acceleration of a body equals the force on it divided by its mass, the knowledge of the forces thus gives us the motion. The x and y coordinates of each body change independently (since the axes are at right-angles) and we need to calculate the components of the forces in the corresponding direction, so that the forces between particles must be resolved into components parallel to x and y. The extension to three dimensions is clear.

The force of gravitational attraction between two particles is $F = G M_1 M_2 r^{-2}$, where G is the gravitational constant, M_1 and M_2 are the masses and r is the distance between them. The components along the x-axis due to each body must be added to give the total force in this direction and similarly for y.

For a stable circular orbit of a smaller mass (1) about a larger mass (2), the centrifugal force equals the gravitational attraction so that $M_1 V^2/r = GM_1 M_2 r^{-2}$. This formula give us the velocity necessary for starting the system in such an orbit. The velocity is brought into the calculation most simply by stating where the body was one interval earlier than the starting time. The period of the circular orbit is $T = 2\pi r/V$. In fact the orbits will be circles about the common centre of gravity, the larger mass moving only slightly as both rotate about the common centre of gravity (see figure 9.1).

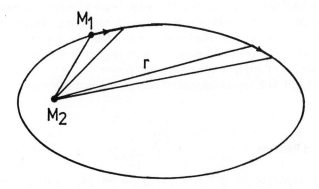

Figure 9.1. The trajectory of a small mass M_1 acted on by the gravitational attraction to a larger mass M_2 is followed in small time intervals by calculating the force acting on the particle and using this to find the new position and velocity at the end of each time interval.

The momentum MV of the body is proportional to the distance between successive positions and it can be changed by adding a vector representing the impulse derived by firing a rocket. This is best applied by changing the previous position by the corresponding vector before calculating the next position from the present position.

We can explore Kepler's Three Laws of Planetary Motion which are: (1) a planet follows an elliptic orbit with the Sun at one focus; (2) the radius vector sweeps out equal areas in equal times; and (3) the squares of the periods are proportional to the cubes of the radii.

However, with three or more bodies, like the Sun, the Earth and the Moon, the behaviour of the system becomes much richer and other phenomena appear. Only a few arrangements are stable. If two bodies come very close together a very large amount of energy may appear in the collision and this may throw the smaller body off the screen at the next step. In fact the mutual gravitational potential is not infinite and we could add a section to allow for this. We may find for very elliptic orbits, where the body has to change direction rapidly, that the time steps we have been taking are too large for accuracy. Various such refinements to the program are possible.

There is considerable scope for experiment in finding out what may happen in a solar system. In order to prevent steady orbits from overlapping on the screen we may give the whole system a small drift velocity to the side. Beginning with two bodies we can make a binary star, or a planet orbiting the Sun, or a comet with too high a velocity to be a satellite, coming in and out again on a hyperbolic path. In the latter case, we could apply velocity corrections by rocket and see if we could put it into circular orbit, like the spacecraft circling the Moon. With three bodies, unexpected things begin to happen and we can try making a Moon, Earth, Sun system or even try navigating a spacecraft among them. We can find out whether three equal Suns circling their common centre of gravity are stable and we can ask what happens if we change from an inverse square law of attraction to some other power.

Program LAPLACE (in the appendix) may provide a starting point.

9.2 Fibonacci Patterns

The Fibonacci series is 1, 1, 2, 3, 5, 8, 13, 21, 34, 55, 89 . . . each number being the sum of the two previous numbers. It has an immense number of properties and there is even a special journal, *The Fibonacci Quarterly*, devoted to them. The ratio of the terms in the series converges to the Golden Number, τ, which is $(1 + \text{sqr}(5))/2 = 1.618 \ldots$.

The Fibonacci series is involved in the patterns of phyllotaxis, for example, the arrangement of seeds in a sunflower head. We can draw such a pattern using the Golden Angle which is $(360/\tau)^\circ = 222.49^\circ$ (or 360° minus this which is

$137.5°$). We draw circles which have polar coordinates with radii $r(i) = k*\text{sqr}(i)$ and $\theta(i) = i*2*\pi/\tau$). Radii increasing as the square root of the integers arrange that the mean density of the packing should be constant and θ increasing in steps of the Golden Angle ensures that the circles are as spread out as possible. The radius of the circles should be about $\tau/2$. (We might note that for the square root of a number some forms of BASIC use SQR(i) and some use SQRT(i)).

Figure 9.2. The Fibonacci spiral is generated by drawing circles with centres at radii increasing as the square roots of the integers and at successive angular intervals of $135.5°$. (Drawing by R. Erickson.)

A program for the Sinclair QL might be: (where we have a screen area 100 x 100 with the point 50,50 in the centre of the screen)

```
100 REMark program name "FIBONACCI"
110 CLS
```

```
120 POINT 50,50
130 t=(1+sqrt(5))/2
140 th=2*PI/t
150 REMark adjustable scale factor
160 k=3
170 rds=k*t/2
180 FOR i=1 to 400
190 r=k*SQRT(i)
200 theta=th*i
210 x=r*SIN(theta)
220 y=r*COS(theta)
225 REM draw circle of radius rds and centre x+50, y+50
230 CIRCLE x+50, y+50, rds
240 NEXT i
```

It would be an interesting problem to see if this technique could be modified to pack circles on the surface of a sphere.

9.3 The Penrose Pattern

A very interesting pattern came suddenly into prominence because of its occurrence in the structure of rapidly cooled alloys. R. Penrose and others had been investigating the construction of jig-saw puzzle patterns, with only two different types pieces which, when joined together, would give a pattern which was non-periodic. For example, pieces which are identical copies of a quadrilateral (convex or concave) automatically pack together to give a 'crystalline' pattern which repeats regularly on a lattice. Penrose found that it was possible to define two tiles which packed together only non-periodically. Another way of generating this pattern is by inflation or deflation. A large tile can be divided into smaller tiles and if this procedure is repeated indefinitely (changing the scale appropriately) we can cover as large an area as we wish. A recursive program (PENROSE) can be written to generate this pattern; see the appendix.

9.4 The Julesz Pattern – A Test for Stereo Perception

Some people, perhaps as many as 10 per cent of the population, find great difficulty in seeing a pair of pictures stereoscopically, that is, in achieving stereopsis.

At Bell Telephone Laboratories, Bela Julesz investigated this ability and, in the *Scientific American* (Julesz, 1965), described an interesting test. In this test, the picture is a random array of small black and white squares making up

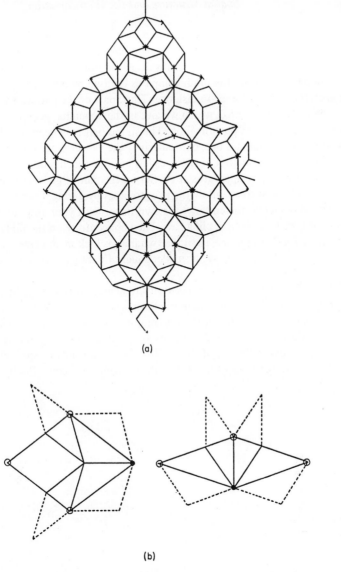

(a)

(b)

Figure 9.3. (a) The two-dimensional Penrose tiling. Given rules for the recursive subdivision of thick and thin rhombi (see (b)) with angles 72° and 144°, this process can be repeated until a pattern with as many tiles as required may be obtained. The pattern is non-periodic, although obtained by this definite rule. (b) The Penrose rule for subdividing the thick and thin tiles to generate an infinite network. Each thick tile becomes two thick and one thin, and each thin tile becomes one thick and one thin in the next generation. A dot marks one end of the tile in one generation and a circle marks corresponding ends in the next.

a larger square. The pattern is generated by a computer and slightly different versions are presented to the left and right eyes of the subject. In the version seen by one eye part of the pattern is displaced horizontally, that is, is given a certain parallax. When the brain compares the two pictures, this parallax is interpreted as a difference in distance. Thus, part of the pattern stands out.

The program JULESZ, which we give in the appendix, generates a pair of pictures showing this phenomenon. The stereo pair may be viewed by any of the standard techniques but, particularly if the viewer is aged more than 50, when the accommodation of the eyes has become much reduced, the effect can be seen without any apparatus at all. Set up the pair of pictures at a distance of 3–10 feet, look at them, allow the eyes to cross by gazing if necessary at a nearer mark such as the end of your own finger with your arm outstretched, until you see three pictures in a row. Then concentrate on the middle one, if necessary for 2–3 minutes, until the two images lock in and give the stereo picture. You should then see a smaller square floating a few inches above the centre of the picture.

The processing in the brain necessary to achieve this percept by correlating corresponding parts of two random pictures is quite remarkable.

To follow the working of the program it is necessary to consult the manual for the printer, to see how dot-images are printed. This program represents a simple example of the use of this facility. Note that the line spacing is changed so that successive scans of the 8 dots in the print head cover the field without gaps.

10 Fourier transforms

Before beginning our mathematical section on the geometry of transform space it is useful first to describe the situation in words.

If a problem is difficult, we try to 'look at it in another way' or to 'turn it round so that we can see it differently'. The techniques of transforming a situation are basic to mathematics and there are very many different kinds of transformation. A picture, a distribution of density in a space of one, two or three dimensions may be transformed and represented in a new way in another space. One of the most important ways is as a Fourier transform (called after Jean Baptiste Joseph Fourier who first described the method in 1812 when dealing with problems in the flow of heat). This analyses a picture or situation in terms of the frequencies which make it up. It is used in everyday life in considering sound or radio waves. When constructing an aerial to receive we have to think of the wavelength, which might be 100 m, and make our aerial half or a quarter of a wavelength long. We imagine an electric wave 100 m from crest to crest, rolling past our aerial and interacting with it. However, when recording the corresponding position on the scale of the radio set, we mark the wave as having a frequency of 3 MHz (meaning 3 million cycles per second). These two measures are connected reciprocally by the velocity of 3×10^8 m/s. A wavelength of one-half of this, 50 m, is marked by a frequency of 6 MHz, twice as far along our frequency scale. The scale of our radio set is then a space in which waves are represented in transformation. A set of marker waves of wavelengths, 1, 2, 4, 8 m, appears reciprocally at distances 1, 0.5, 0.25, 0.125 along the scale which might be called 'reciprocal space'.

We may set out quantitatively the properties of the Fourier transform as follows:

(1) A density distribution can be built up by the linear superposition of a number of sine waves of different frequencies. If the density distribution is periodic, of frequency f (like the complex note of a violin), then these sine waves have only the frequencies $f, 2f, 3f \ldots nf \ldots$ but if the density distribution is non-repeating, like a single pulse, then an infinite band of frequencies is required. Resolving a complex signal into its Fourier components (the sine waves) is known as Fourier analysis. Combining the sine waves together again to give the general signal is correspondingly called Fourier synthesis. Both processes were formerly very time-consuming if manual methods were used, but the situation has been totally changed by

83

the advent of the computer. At first a main-frame computer was necessary, but now the methods are passing into the repertoire of the microcomputer.

(2) For each constituent sine wave we have to state its frequency, its amplitude and its phase (where the peak falls with respect to some reference signal). If the wave is in more than one dimension then we must also state its direction. A wave in space is represented by the direction of the normal to the wavefronts. As we will see, complex numbers provide an appropriate mathematical representation of amplitude and phase.

(3) In many physical systems, such as for radio waves in space or for light in a camera, the principle of linear superposition obtains. This means that wave trains can cross each other without getting mixed up. If linear superposition does not apply then two waves cross-modulate each other and new sum and difference frequencies arise. Non-linear devices have their important place. If we have moving waves then it is vital to know whether waves of all frequencies travel with the same velocity (as is the case for electromagnetic waves in free space, but is not so for a medium such as glass). We have to be clear on the properties of the particular system with which we are dealing.

(4) The spectrum of sine waves which make up a particular density distribution is known as its *Fourier transform*. In representing the transform we have to make provision for noting the amplitude, frequency, phase and direction of every constituent wave. We can plot out the transform in *transform space*. If the scales used in the *direct space* or *real space* in which we plot our original density distribution are, for example, centimetres, then the scales used in the transform space will be reciprocal, for example, waves per centimetre, which we may call reciprocal centimetres. Using the same word 'centimetres' may cause confusion but, in dealing with atomic structures we have a unit, the Ångstrom unit, which is 10^{-8} cm. In plotting out our density distribution we may use a scale of so many centimetres to represent each Ångstrom unit. Then when plotting the transform we use another scale of so many centimetres on the paper to represent each *reciprocal Ångstrom unit*. A three-dimensional density distribution has a three-dimensional Fourier transform.

(5) When we buy an audio amplifier we ask about its frequency response and hope that all frequencies of notes in the complex sound signal going in will come out 'undistorted'. The curve which describes the response of the system to different frequencies is called the 'contrast transfer function'. Asking about the contrast transfer function of an optical lens or of an electron microscope or of some other system has revolutionised our understanding of these instruments.

Many problems, then, are simplified if we first resolve a complex density distribution into sine waves and ask what happens to each sine wave separately as it passes through the system and then re-assemble the somewhat changed sine waves to give a resultant density distribution. What happens to each sine wave may be represented in transform space and is the contrast transfer function mentioned above. Waves may be changed in amplitude, phase or direction. We may be considering the properties of a physical system, such as a microscope, or we may be using a computer as a kind of generalised imaging system, where we can apply any contrast transfer function, physically realisable or not.

(6) We have to build up a kind of library or dictionary of typical density distributions and their transforms. Such lists, in mathematical form, are to be found in handbooks. Visually, a picture in real space corresponds to another picture in transform space. It is important to acquire a qualitative pictorial appreciation of a range of such relationships to understand what is happening mathematically. The dictionary is reciprocal and works equally in both directions — transforming a transform takes us back to the original distribution.

(7) There is an important theorem or principle, the *convolution theorem*, which helps us to build up this library. It tells us how to think about the transforms of complex patterns in terms of the transforms of simpler objects. The concept of *convolution* is required. If we have two density distributions A and B (two pictures, for example), then the convolution of A and B means repeating the whole of A at every point of B, or vice versa (it does not matter which way we go). For example, if A is a circle and B is a lattice of points, then the convolution of A and B (which we may write as conv(A, B)), is a lattice of circles. Physically, A might be a molecule of naphthalene and B might be a three-dimensional lattice of points. The convolution of A and B would then be a crystal of molecules of naphthalene stacked in a three-dimensional array.

The convolution theorem then states that if we have a distribution in real space which is the point by point product of the distributions A and B, then the transform of this distribution is the convolution of transform(A) with transform(B). The theorem works in either direction. If in real space the distribution C is the convolution of the distributions A and B, the transform(C) is the point by point product of the transform of A and the transform of B.

(8) The convolution theorem enables us to do many remarkable things. During the Apollo 13 space flight the astronauts took a photograph of their damaged spacecraft, but it was out of focus. It is possible by image processing methods to put such an out-of-focus picture back into focus and thus to clarify it. It is simpler to take the example of a photograph of a person which has been doubly exposed. Two overlapping images, separated for example by 1 mm appear in the print. How can a clear picture be obtained? Suppose this picture C is the convolution of two separate distributions A and B. A is the wanted picture, the image of one person and B is a distribution of two points only. C is obtained by repeating A at every point of B. The transform of C can be obtained by calculation. The trans-

form of B (the two points) is simple and is known. Since $C = \text{conv}(A, B)$ we have $\text{trans}(C) = \text{trans}(A) \times \text{trans}(B)$. $\text{trans}(B)$ can then be obtained by point-by-point division of $\text{trans}(C)$ by $\text{trans}(B)$. It is then re-transformed and $\text{trans}(\text{trans}(A)) = A$ is obtained. This process is known as deconvolution and is easily accomplished by transforming our initial problem into Fourier transform space. This can be done by optical analogue methods or by purely computational methods, the latter having now largely superseded the former.

Before formulating the mathematics of the Fourier transform an example is given to illustrate the power of Fourier processing methods applied to the analysis of biological density distributions.

One of the most widely used techniques for the study of biological materials, such as viruses or components of isolated cells, is electron microscopy. This process involves exposing the biological material to a beam of impinging electrons which are then scattered by the material. The subsequent pattern of the scattered electrons is recorded photographically and it is this pattern which provides details about the structure of the material. The photograph is referred to as an electron micrograph and can be automatically converted into a set of digital density values by assigning numbers to the different levels of intensity within the micrograph. Figure 10.1a shows an electron micrograph of material comprising one of the surface proteins which forms part of the structure of the influenza virus. Figure 10.1b shows an enlarged display of the digitised micrograph. This digitised set of densities is fed into the computer, the Fourier transform applied and the corresponding frequency components displayed (figure 10.1c). The figure, however, shows two types of distribution; one which consists of reasonably defined spots with a regular spacing between them, and the other which consists of smaller spots with no apparent regular spatial features. The former set of spots correspond to the basic frequencies of the material and the latter arise from the 'noise' components which occur during the process of obtaining the electron micrograph.

The 'noise' constitutes an unwanted part of the micrograph, since it obscures much of the structural detail and we must therefore try and remove it. We achieve this by retaining only those spots in the frequency display which correspond to the basic structural frequencies and then apply the inverse Fourier transform (see section 10.7) to arrive at a 'noise free' result. Figures 10.1d and 10.1e show the result of this operation. Figure 10.1e corresponds to the area shown in 10.1b, and figure 10.1d corresponds to a computer-contoured display for a selected area of the 'noise free' image. The most striking feature of this process, called digital spatial filtering, can be seen in the comparison between figures 10.1b and 10.1e, where the structural details of the image can be clearly seen, whereas they were previously obscured by 'noise'.

The calculations were all carried out on a main-frame computer, because of the size of the density distributions being processed (approximately 250 000 points). As yet, it is not possible to apply the above procedure to data sets of

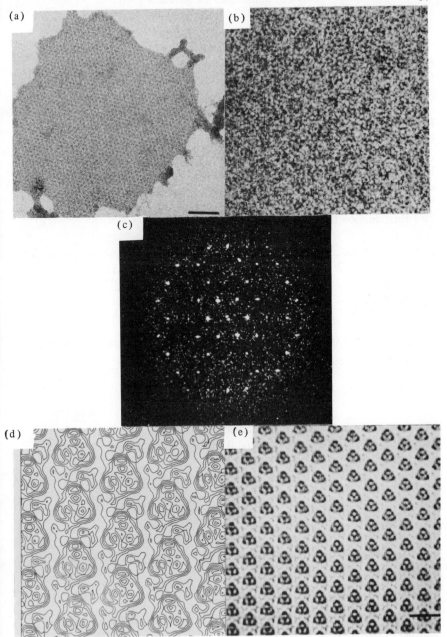

Figure 10.1. (a) Electron micrograph of protein material from influenza virus.
(b) Display of digitised area within micrograph. (c) Computerised Fourier
transform of (b). (d) Contoured display of selected area from 'noise free' image.
(e) Display of 'noise free' image corresponding to area shown in (b). (Pictures
courtesy of E. B. Brown.)

this size on a microcomputer, since the memory and disc capacities are too limited. Nevertheless, procedures of this type can be carried out on much smaller data sets using a microcomputer and it is with this use in mind that the subsequent sections explain in precise mathematical terms how these processes are carried out.

The rest of this chapter discusses the fundamental theory of Fourier transforms applied to both continuous and discrete data sets. Particular attention is paid to the FFT (fast Fourier transform) as well as to other methods which optimise the organisation of the data in order to optimise processing times. Each program is provided with documentation and an example illustrating the mode of input, together with the corresponding output.

10.1 Introduction to the Fourier Transform

If f(x) is a continuous function* of the real variable x the Fourier transform of f(x), denoted by $\mathfrak{F}\{f(x)\}$ is defined by the equation (where $i = \sqrt{-1}$)

$$\mathfrak{F}\{f(x)\} = F(u) = \int_{-\infty}^{+\infty} f(x)\exp(-2\pi iux)\,dx \tag{10.1}$$

Conversely given $F(u)$, f(u) can be obtained by using the inverse Fourier transform defined as

$$\mathfrak{F}^{-1}\{F(u)\} = f(x) = \int_{-\infty}^{+\infty} F(u)\exp(2\pi iux)\,du \tag{10.2}$$

The above two functions form what is called a transform pair, and one is the complex conjugate of the other — that is, the sign of the exponent argument is reversed.

Thus the Fourier transform of sin 2x would be

$$\mathfrak{F}\{\sin 2x\} = F(u) = \int_{-\infty}^{+\infty} \sin 2x \exp(-2\pi iux)\,dx$$

The expression $\exp(-2\pi iux)$ can be simply re-written in a complex number notation as

$$\exp(-2\pi iux) = \cos 2\pi ux - i\sin 2\pi ux$$

using Euler's formula which states that $\exp(i\theta)$ may be expressed in the form $\exp(i\theta) = \cos\theta + i\sin\theta$.

---

*Continuous functions can be briefly described as those which exist for all chosen values of the function variable. Examples are sin 2x and cos 2x. However, it should be noted that a function does not have to be continuous in order for its Fourier transform to be determined (see section 10.2).

Thus, although our function $f(x)$ may be a real function, the resulting transforms are generally complex — that is, expressible in the form $R(u) + iI(u)$ where $R(u)$ and $I(u)$ are the real and imaginary parts of the resulting transform.

As an example, consider the function for which $f(x) = 1$ for $0 \leqslant x \leqslant 1$, $f(x) = 0$ elsewhere (figure 10.2).

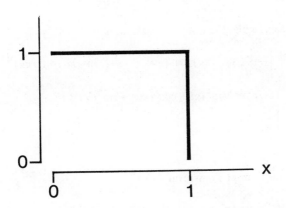

Figure 10.2

Then
$$F(u) = \int_{-\infty}^{+\infty} f(x)\exp(-2\pi iux)\, dx$$

$$= \int_{0}^{1} \exp(-2\pi iux)\, dx$$

$$= -(1/2\pi iu)[\exp(-2\pi iux)]_{0}^{1} = -(1/2\pi iu)\exp(-2\pi iu - 1)$$
$$= +(1/2\pi i)[\exp(i\pi u) - \exp(-i\pi u)]\exp(-i\pi u) =$$
$$= (1/\pi u)\sin(\pi u)\exp(-i\pi u)$$

This is a complex function for which the real part $R(u)$ is given by

$$R(u) = \frac{1}{\pi u}\sin(\pi u)\cos(\pi u)$$

The corresponding imaginary part $I(u)$ is given by

$$I(u) = -\frac{1}{\pi u}\sin^2(\pi u)$$

It is frequently important to examine the Fourier amplitude or magnitude $|F(u)|$ of $f(x)$ and its associated phase angle $\phi(u)$. These are defined as

$$F(u) = |F(u)| \exp[i\phi(u)]$$

where $|F(u)| = [(R(u))^2 + (I(u))^2]^{\frac{1}{2}}$

$$\phi(u) = \tan^{-1}[I(u)/R(u)]$$

The range of values of $|F(u)|$ defines what is known as the Fourier amplitude spectrum.

In the above example, the Fourier spectrum is

$$|F(u)| = \left| \left[\frac{1}{\pi^2 u^2} \sin^2(\pi u)\cos^2(\pi u) + \frac{1}{\pi^2 u^2} \sin^2(\pi u)\sin^2(\pi u) \right] \right|^{\frac{1}{2}}$$

$$= |(1/\pi^2 u^2)\sin^2(\pi u)\{\cos^2(\pi u) + \sin^2(\pi u)\}|^{\frac{1}{2}}$$

$$= \left| \frac{1}{\pi u} \sin(\pi u) \right|$$

10.2 The Discrete Fourier Transform

Figure 10.3 shows a continuous function $f(x)$ sampled at points which are Δx units apart. The continuous function now takes the form

$$f(x_0), f(x_0 + \Delta x), f(x_0 + 2\Delta x), f(x_0 + 3\Delta x), f(x_0 + 4\Delta x), \ldots f(x_0 + n\Delta x)$$

that is

$$f(x) = f(x_0 + x\Delta x)$$

where x assumes the discrete values $0, 1, 2, \ldots, N - 1$.

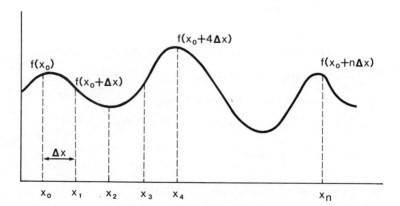

Figure 10.3. Continuous function sampled at points Δx units apart.

The Fourier transform of this discrete function may be written as

$$F(u) = \frac{1}{N} \sum_{x=0}^{N-1} f(x)\exp(-2\pi iux/N)$$ (10.3)

for $u = 0, 1, 2, \ldots, N-1$ and conversely

$$f(x) = \sum_{u=0}^{N-1} F(u)\exp(+2\pi iux/N)$$ (10.4)

for $x = 0, 1, 2, \ldots N-1$.

The corresponding Fourier transform in two-dimensional space may be written as

$$F(u, v) = \frac{1}{MN} \sum_{x=0}^{N-1} \sum_{y=0}^{N-1} f(x,y)\exp[-2\pi i(ux/M + vy/N)]$$

for $u = 0, 1, 2, \ldots, M-1; v = 0, 1, 2, \ldots, N-1$

and conversely

$$f(x,y) = \sum_{u=0}^{M-1} \sum_{v=1}^{N-1} F(u, v)\exp[2\pi i(ux/M + vy/N)]$$

for $x = 0, 1, 2, \ldots, M-1; y = 0, 1, 2, \ldots, N-1$.

As shown above, sampling a function in one dimension gives a set of points $f(x + \Delta x)$. Sampling a two-dimensional function similarly gives a set of points $f(x + \Delta x, y + \Delta y)$, that is, a two-dimensional rectangular grid of points Δx units apart in one direction and Δy units apart in the other.

10.3 Computing the One-dimensional Fourier Transform

Figure 10.4 shows a continuous function sampled at four points whose x co-ordinate values are 0; 1; 2; 3. The corresponding values for $f(x)$ are 1; 2; 3; 4.

Application of eqn (10.3) to determine the Fourier transform of $f(x)$ sampled at the above points is as follows

$$
\begin{aligned}
F(0) &= \frac{1}{4} \sum_{x=0}^{3} f(x)\exp(0) &= \frac{1}{4}[f(0) + f(1) + f(2) + f(3)] \\
&= \frac{1}{4}[1 + 2 + 3 + 4] &= 2.5
\end{aligned}
$$

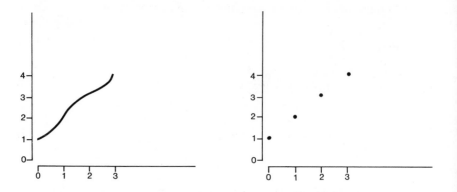

Figure 10.4. Continuous function sampled at four points.

$$F(1) = \frac{1}{4}\sum_{x=0}^{3} f(x)\exp(-2\pi ix/4)$$

$$= \frac{1}{4}[1\exp(0) + 2\exp(-\pi i/2) + 3\exp(-\pi i) + 4\exp(-3\pi i/2)]$$

$$= \frac{1}{4}[-2 + 2i]$$

$$F(2) = \frac{1}{4}\sum_{x=0}^{3} f(x)\exp(-4\pi ix/4)$$

$$= \frac{1}{4}[1\exp(0) + 2\exp(-\pi i) + 3\exp(-2\pi i) + 4\exp(-3\pi i)]$$

$$= \frac{1}{4}[-2 + 0i]$$

$$F(3) = \frac{1}{4}\sum_{x=0}^{3} f(x)\exp(-6\pi ix/4)$$

$$= \frac{1}{4}[1\exp(0) + 2\exp(-3\pi i/2) + 3\exp(-3\pi i) + 4\exp(-9\pi i/2)]$$

$$= \frac{1}{4}[-2 - 2i]$$

The corresponding Fourier amplitude spectrum $|f(u)|$ and associated phase $\phi(u)$ are calculated as

$|F(0)| = 2.5$

$\phi(0) = \tan^{-1}[0] = 0$

$|F(1)| = [(-\tfrac{1}{2})^2 + (\tfrac{1}{2})^2]^{\frac{1}{2}} = 0.707$

$\phi(1) = \tan^{-1}[-1] = -45°$

$|F(2)| = [(-\tfrac{1}{2})^2 + (0)^2]^{\frac{1}{2}} = 0.5$

$\phi(2) = \tan^{-1}[0] = 0°$

$|F(3)| = [(-\tfrac{1}{2})^2 + (-\tfrac{1}{2})^{\frac{1}{2}}] = 0.707$

$\phi(3) = \tan^{-1}[1] = 45°$

Figure 10.5 shows the amplitude and corresponding phase spectrum.

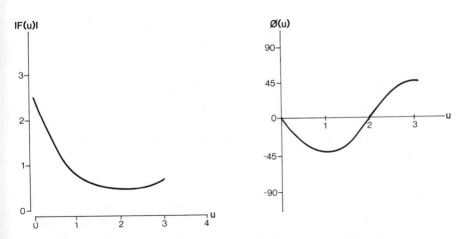

Figure 10.5. Resulting Fourier transform amplitude and phase spectrum for sampled function.

The program ONEDFT is provided for calculating the discrete Fourier transform (using eqn (10.3)) and its associated amplitude and phase spectrum for a set of equally sampled points, and is illustrated in relation to the above example.

Returning to figure 10.4, it is evident that we have not reconstructed the function $f(x)$ accurately since our sampling is too crude. Clearly the more points we use to sample the function the more precisely we are able to obtain an accurate representation of the function.

Figure 10.6 shows the effect of sampling the same function at about twice as many points as previously. Our reconstruction of the original function is clearly more representative. This raises two important questions in relation to the calculation of the Fourier transform of a discrete function:

1. How is the Fourier transform affected by including more points?
2. How many points are necessary to calculate the Fourier transform accurately?

In a sense both of these questions are related, but merely to illustrate the former question we return to our example and show the effect of incorporating the additionally sampled values.

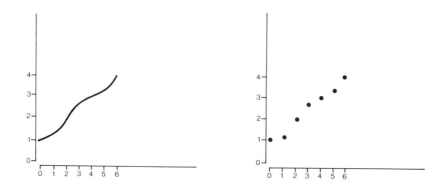

Figure 10.6. Previous function sampled at more points.

Running the program ONEDFT for the seven points whose values correspond to 1.0, 1.25, 2.0, 2.81, 3.0, 3.44, 4.0 gave the amplitude and phase spectrum shown in figure 10.7. This should be compared with the corresponding phase and amplitude spectrum shown in figure 10.5. The results clearly show differences which reflect the effect of under-sampling the function, particularly in the phase spectrum. We could argue therefore that we should include as many points as possible in order to achieve accuracy.

There are two problems with this argument however. In the first case, since our points are often determined as data values from a particular experimental source, the limitations imposed by the experimental techniques involved may restrict us to a limited range of data points. In the second case, the Fourier transform of a particular function can be optimally determined for a sub-set of data values depending on the nature of the function and the accuracy with which we require to evaluate the transform. Consideration of these points leads to a discussion of the problems of optimal sampling and resolution, as discussed in the following section.

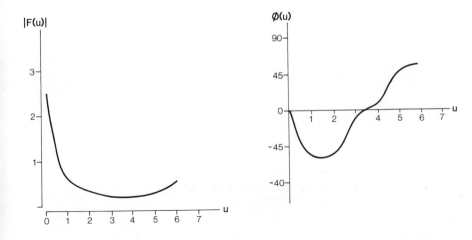

Figure 10.7. Corresponding Fourier transform and phase spectrum.

10.4 Sampling and Resolution

Consider the function shown in figure 10.8, assumed to extend from $-\infty$ to $+\infty$, and suppose that the Fourier transform of $f(x)$ vanishes for values of u outside the interval $[-W, W]$. A function whose transform has this property for any finite value of W is called a band-limited function. To recover this function completely requires that the sampling Δx is chosen so that

$$\Delta x \leqslant \frac{1}{2W}$$

This result is referred to basically as the Whittaker–Shannon sampling theorem. This sampling requirement is a strictly theoretical one and in practice it may be necessary to sample the data even more finely in order to recover the function. This situation occurs because the function or signal, as it is more commonly called, may be obscured by distortive effects, particularly when it originates from an experimental source.

As a further illustration of the Fourier transform, three distributions are calculated using the program GENDAT and their corresponding amplitude spectra illustrated in figure 10.9a. The distributions chosen were the Gaussian, Poisson and Sine distributions. A continuous curve has been drawn through the points used as input to the transform. The amplitude spectra for these distributions will generally differ slightly from those shown in many of the standard textbooks, particularly in the case of the sine curve. The reason for this is that the functions we have chosen are evaluated over a limited number of points (discrete sample) and over a finite level. Those given in standard textbooks are

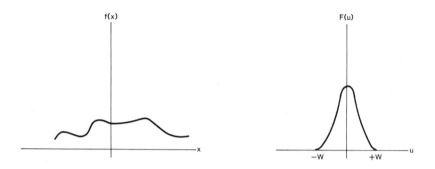

Figure 10.8. Examples of a band-limited function.

evaluated as continuous functions between $\pm \infty$. Figure 10.9b shows the effect on the Fourier amplitude spectra when the sampling of the data is halved — that is, every other point is selected for input into the transform. The results show that for *this* particular case the effect of halving the sampling interval makes little change to the basic profiles of the amplitude spectra.

10.5 The Fast Fourier Transform

Consider the evaluation of eqn (10.3) for the N values of u corresponding to $u = 0, 1, 2, \ldots, (N-1)$

$$F(0) = [f(0)\exp(-2\pi i\; 0 \times 0/N) + f(1)\exp(-2\pi i\; 0 \times 1/N) + \ldots$$

$$F(1) = [f(0)\exp(-2\pi i\; 1 \times 0/N) + f(1)\exp(-2\pi i\; 1 \times 1/N) + \ldots$$

.

.

.

$$F(N-1) = [f(0)\exp\{-2\pi i(N-1) \times 0/N\} + f(1)\exp\{-2\pi i(N-1) \times 1/N\} + \ldots$$

Evaluation of F(0) requires a total of $(2N-1)$ operations since we have to perform N multiplications (that is, $f(x) \times \exp \ldots$) and $(N-1)$ additions. Similarly the evaluation of F(1) requires $(2N-1)$ operations and so on until the evaluation of $F(u)$ over all N values of u is completed. This requires a total number of operations proportional to N^2. (Note that the evaluation of $\exp(-2\pi iux/N)$ for values of u and x can be computed once and stored in a table for subsequent operations, and does not therefore constitute a direct part of the implementation.)

We show in this section that by algebraically decomposing eqn (10.3) the number of operations can be reduced from N dependence to $N\log_2 N$, subject to N being expressed as a power of 2.

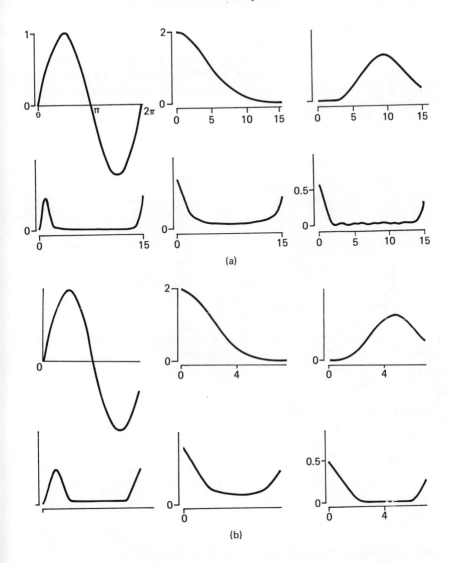

Figure 10.9. (a) Top shows curves corresponding to Sine, Gaussian and Poisson distributions. Bottom shows the respective Fourier transform amplitude spectra. (b) Same distribution as previous but with only half the number of values.

At first sight this does not appear to be a very dramatic saving in terms of the number of operations required. If, however, we consider the evaluation of eqn (10.3) for $N = 2^{13}$ or 8192 data points the number of operations is reduced from 67 108 864 (N^2) to 106 496 ($N\log_2 N$); that is, a saving by a factor of 67 in the number of operations.

The computational advantages in reducing the number of operations this way results in saving of many hours of CPU time for practical purposes since we often require to evaluate a Fourier transform over as many as a quarter of a million or more data points. Table 10.1 shows a table of N^2 against $N\log_2 N$ for values of N up to 2^8. We can clearly see that as N increases the effective number of operations is significantly reduced.

Table 10.1

N	N^2	$N\log_2 N$
2	4	2
4	16	8
8	64	24
16	256	64
32	1024	160
64	4096	384
128	16384	896
256	65536	2048

FFT (fast Fourier transform) algorithm

The FFT algorithm developed in this section is based on the so-called 'successive doubling' method. It will be convenient in the following discussion to express eqn (10.3) in the form

$$F(u) = \frac{1}{N} \sum_{x=0}^{N-1} f(x) W_N^{ux} \tag{10.5}$$

where

$$W_N = \exp(-2\pi i/N) \tag{10.6}$$

and N is assumed to be of the form

$$N = 2^n \tag{10.7}$$

where n is a positive integer. Based on this, N can be expressed as

$$N = 2M \tag{10.8}$$

where M is also a positive integer. Substitution for N in eqn (10.5) yields

$$F(u) = \frac{1}{2M} \sum_{x=0}^{2M-1} f(x) W_{2M}^{ux}$$

$$= \frac{1}{2} \frac{1}{M} \sum_{x=0}^{M-1} f(2x) W_{2M}^{u}(2x) + \frac{1}{M} \sum_{x=0}^{M-1} f(2x+1) W_{2M}^{u}(2x+1) \tag{10.9}$$

Since, from eqn (10.6) $W_{2M}^{2ux} = W_M^{ux}$, eqn (10.9) may be expressed in the form

$$F(u) = \frac{1}{2}\,\frac{1}{M}\sum_{x=0}^{M-1} f(2x)W_M^{ux} + \frac{1}{M}\sum_{x=0}^{M-1} f(2x+1)W_M^{ux}W_{2M}^{u} \qquad (10.10)$$

If we define

$$F_{\text{even}}(u) = \frac{1}{M}\sum_{x=0}^{M-1} f(2x)W_M^{ux} \qquad (10.11)$$

for $u = 0, 1, 2, \ldots, M-1$ and

$$F_{\text{odd}}(u) = \frac{1}{M}\sum_{x=0}^{M-1} f(2x+1)W_M^{ux} \qquad (10.12)$$

for $u = 0, 1, 2, \ldots M-1$, eqn (10.10) then becomes

$$F(u) = \tfrac{1}{2}\{ F_{\text{even}}(u) + F_{\text{odd}}(u)W_{2M}^{u}\} \qquad (10.13)$$

Also, since W_M^{u+M} and $W_{2M}^{u+M} = -W_{2M}^{u}$, it follows from eqns (10.11)–(10.13) that

$$F(u+M) = \tfrac{1}{2}\{ F_{\text{even}}(u) - F_{\text{odd}}(u)W_{2M}^{u}\} \qquad (10.14)$$

Careful analysis of eqns (10.11)–(10.14) shows that N point transforms can be computed by dividing the original expression into two parts, as indicated in eqns (10.13) and (10.14). Computation of the first half of $F(u)$ requires the evaluation of the two $(N/2)$ point transforms given in eqns (10.11) and (10.12). The resulting values of $F_{\text{even}}(u)$ and $F_{\text{odd}}(u)$ are then substituted into eqn (10.13) to obtain $F(u)$ for $u = 0, 1, 2, \ldots (N/2 - 1)$. The other half then follows directly from eqn (10.14) without additional transform evaluations.

In order to examine the computational implications of the above procedure, let $m(n)$ and $a(n)$ represent the number of complex multiplications and additions respectively required to implement this method. As before, the number of samples is equal to 2^n where n is a positive integer. Suppose first that $n = 1$. A two-point transform requires the evaluation of $F(0)$; then $F(1)$ follows from eqn (10.14). To obtain $F(0)$, we must first compute $F_{\text{even}}(0)$ and $F_{\text{odd}}(0)$. In this case $M = 1$ and eqns (10.11) and (10.12) are one-point transforms. Since the Fourier transform of a single point is the sample itself, however, no multiplications or additions are required to obtain $F_{\text{even}}(0)$ and $F_{\text{odd}}(0)$. One multiplication of $F_{\text{odd}}(0)$ by W_2^0 and one addition yield $F(0)$ from eqn (10.13). $F(1)$ then follows from eqn (10.14) with one more addition (we consider subtraction and addition as synonymous). Since $F_{\text{odd}}(0)\,W_2^0$ was already computed, we have that the total number of operations required for a two-point transform consists of $m(1) = 1$ multiplications and $a(1) = 2$ additions.

The next allowed value for n is 2. According to the above development, a four-point transform can be divided into two parts. The first half of $F(u)$ requires evaluation of two, two-point transforms, as given in eqns (10.11) and (10.12) for $M = 2$. Since a two-point transform requires $m(1)$ and $a(1)$ multiplications and additions respectively, it is evident that evaluation of these two equations requires a total of $2m(1)$ multiplications and $2a(1)$ additions. Two additional multiplications and two additions are necessary to obtain F(0) and F(1) from eqn (10.13). Since $F_{odd}(u)W_{2M}^u$ was already computed for $u = \{0, 1\}$, we have that two more additions give F(2) and F(3). The total is then $m(2) = 2m(1) + 2$ and $a(2) = 2a(1) + 4$.

When n is equal to 3, we consider two, four-point transforms in the evaluation of $F_{even}(u)$ and $F_{odd}(u)$. These require $2m(2)$ multiplications and $2a(2)$ additions. Four more multiplications and eight additions yield the complete transform. The total is then $m(3) = 2m(2) + 4$ and $a(3) = 2a(2) + 8$.

By continuing this argument one would find that, for any positive integer value of n, the number of multiplications and additions required to implement the FFT is given by the recursive expressions

$$m(n) = 2m(n-1) + 2^{n-1} \qquad\qquad n \geqslant 1 \qquad (10.15)$$

$$a(n) = 2a(n-1) + 2^n \qquad\qquad n \geqslant 1 \qquad (10.16)$$

where $m(0) = 0$ and $a(0) = 0$, since the transform of a single point does not require any additions or multiplications.

Implementation of eqns (10.11) and (10.12) constitutes the successive doubling FFT algorithm. The name arises from the fact that a two-point transform is computed from two, one-point transforms, a four-point transform from two, two-point transforms, and so on for any N that is equal to an integer power of 2.

Number of operations

It is shown by induction that the number of complex multiplications and additions required to implement the above FFT algorithm is given by

$$m(n) = \tfrac{1}{2}2^n\log_2 2^n$$
$$= \tfrac{1}{2}N\log_2 N$$
$$= \tfrac{1}{2}Nn \qquad\qquad n \geqslant 1 \qquad (10.17)$$

and

$$a(n) = 2^n\log_2 n$$
$$= N\log_2 N$$
$$= Nn \qquad\qquad n \geqslant 1 \qquad (10.18)$$

First it is necessary to prove that eqns (10.15) and (10.16) hold for $n = 1$. It has already been shown that

$$m(1) = \tfrac{1}{2}(2)(1) = 1$$

and

$$a(1) = (2)(1) = 2$$

Next, it is assumed that the expressions hold for n. It is then required to prove that they are also true for $n + 1$. From eqn (10.15) it follows that

$$m(n + 1) = 2m(n) + 2^n$$

Substituting eqn (10.17) which is assumed to be valid for n, yields

$$
\begin{aligned}
m(n + 1) &= 2(\tfrac{1}{2}Nn) + 2^n \\
&= 2(\tfrac{1}{2}2^n n) + 2^n \\
&= 2^n(n + 1) \\
&= \tfrac{1}{2}2^{n+1}(n + 1)
\end{aligned}
$$

Equation (10.17) is, therefore, valid for all positive integer values of n. From eqn (10.16), we have

$$a(n + 1) = 2a(n) + 2^{n+1}$$

Substitution of eqn (10.18) for $a(n)$ yields

$$
\begin{aligned}
a(n + 1) &= 2Nn + 2^{n+1} \\
&= 2(2^n n) + 2^{n+1} \\
&= 2^{n+1}(n + 1)
\end{aligned}
$$

thus completing the proof.

Programming the FFT

The programming method for the FFT relies on successive application of eqns (10.11) and (10.12). Before implementing this, however, we have to re-order the data. Two examples will illustrate the re-ordering procedure.

Consider the array with function values $f(0), f(1), f(2), f(3), f(4), f(5)$, $f(6), f(7)$. This array has 8 or 2^3 values and we are required to determine its Fourier transform using the FFT algorithm based upon successive doubling. Equation (10.11) uses the samples with even arguments — that is, $f(0), f(2)$, $f(4), f(6)$ — and eqn (10.12) uses the samples with odd arguments — $f(1)$, $f(3), f(5), f(7)$. Each of the above arrays can be further re-arranged into odd and even parts, however, since the even parts for the first array are $f(0), f(4)$ while the odd parts are $f(2), f(6)$. Similarly, the even and odd parts for the second array are f(1), f(5) and f(3), f(7) respectively. No further re-arrangement is necessary since each of the above pairs can be considered as consisting of one even part and one odd part; that is, for the pair f(0), f(4) the even part is the term f(0) and the odd part f(4). Equations (10.11)–(10.14) are then applied to the four respective pairs of the re-ordered array to give the required transform.

As a further illustration of the re-ordering process, we consider the 16 function values

$$f(0), \ldots \ldots f(15)$$

which we denote by its arguments

0, 1, 2, 3, 4, 5, 6, 7, 8, 9, 10, 11, 12, 13, 14, 15

The first step in the re-ordering process gives the re-ordering as

0 2 4 6 8 10 12 14 ┊ 1 3 5 7 9 11 13 15

The second successive step gives

0 4 8 12 ┊ 2 6 10 14 ┊ 1 5 9 13 ┊ 3 7 11 15

The final step gives

0 8 ┊ 4 12 ┊ 2 10 ┊ 6 14 ┊ 1 9 ┊ 5 13 ┊ 3 11 ┊ 7 15

The broken lines illustrate where the successive doubling procedure occurs. Thus, eqns (10.11)–(10.14) can be applied to the eight above pairs to give the transform.

If we now return to the array of eight values and consider the original array in relation to the re-ordered array, we have

f(0), f(1), f(2), f(3), f(4), f(5), f(6), f(7) for the original array

f(0) f(4) f(2) f(6) f(1) f(5) f(3) f(7) for the re-ordered array

and we can see that the re-ordering process has involved an interchange between the pairs f(1) and f(4); f(3) and f(6), with the other values remaining unchanged. Now consider the arguments (that is, 0, 1, 2, etc.) of the above arrays expressed as bit patterns.

Argument for original array	Bit representation		
0	0	0	0
1	0	0	1
2	0	1	0
3	0	1	1
4	1	0	0
5	1	0	1
6	1	1	0
7	1	1	1

Argument for re-ordered array	Bit representation
0	0 0 0
4	1 0 0
2	0 1 0
6	1 1 0
1	0 0 1
5	1 0 1
3	0 1 1
7	1 1 1

A comparison between the above two bit patterns shows that one is the reverse of the other; that is, 100 is the reverse of 001 and 110 is the reverse of 011. Correspondingly, our 16-value array would be represented within a total of 4 bits and arguments such as 1, 2, 3 etc. represented in bit form as 0001, 0010, 0011 etc. would, in reverse notation, become 1000, 0100, 1100 etc. These reverse bit patterns correspond to argument values 8, 4, 12 etc. Thus, we can see that, by merely reversing the bit pattern for a given argument value, we obtain its equivalent value in the re-ordering process for the data as input to the FFT. This process of bit reversal is used as a means of the data re-ordering process in most FFT algorithms.

The program FFT is provided for calculating Fourier transforms using the above approach and is illustrated in relation to computing the transform of a square wave. The results are shown in figure 10.10.

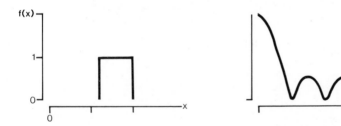

Figure 10.10. Fourier transform of a square wave.

10.6 Separability (the Two-dimensional Transform of a Square Array)

The discrete Fourier transform pair

$$F(u, v) = \frac{1}{N^2} \sum_{x=0}^{N-1} \sum_{y=0}^{N-1} f(x, y) \exp[(-2\pi i/N)(ux + vy)]$$

for $u, v = 0, 1, 2, \ldots, N - 1$ and

$$f(x, y) = \frac{1}{N^2} \sum_{u=0}^{N-1} \sum_{v=0}^{N-1} F(u, v) \exp \left[(+ 2\pi i/N) (ux + vy) \right]$$

for $x, y = 0, 1, 2, \ldots, N - 1$

can be expressed in the separable forms

$$F(u, v) = \frac{1}{N^2} \sum_{x=0}^{N-1} \exp(-2\pi iux/N) \sum_{y=0}^{N-1} f(x, y) \exp(-2\pi ivy/N) \qquad (10.19)$$

for $u, v = 0, 1, 2, \ldots, N - 1$ and

$$f(x, y) = \frac{1}{N^2} \sum_{u=0}^{N-1} \exp(2\pi iux/N) \sum_{v=0}^{N-1} F(u, v) \exp(2\pi ivy/N) \qquad (10.20)$$

for $x, y = 0, 1, 2, \ldots, N - 1$.

The computational significance in expressing our two-dimensional transform in this way is that $F(u, v)$ or $f(x, y)$ can be obtained in two steps by successive applications of the one-dimensional Fourier transform or its inverse.

Equation (10.19) may be expressed in the form

$$F(u, v) = \frac{1}{N^2} \sum_{x=0}^{N-1} F(x, v) \exp(-2\pi iux/N) \qquad (10.21)$$

where

$$F(x, v) = \sum_{y=0}^{N-1} f(x, y) \exp(-2\pi ivy/N) \qquad (10.22)$$

For each value of x, eqn (10.22) represents N times the one-dimensional transform with frequency values $v = 0, 1, \ldots, N - 1$. Therefore, the two-dimensional function $F(x, v)$ is obtained by taking a transform along each *row* of $f(x, y)$ and multiplying the result by N. The result, $F(u, v)$, can then be obtained by taking the transform along each *column* of $F(x, v)$ as indicated by eqn (10.21).

The procedure is illustrated in relation to the following example using both the slow two-dimensional transform TWODFT and the fast transform FFT.

Our array consists of a square containing real values as follows

1	0	0	1
0	0	0	0
1	0	0	1
0	0	0	0

Since we have to submit our values to FFT in pairs where the second value consists of the imaginary part of the number (in this case, zero), we express the above array in terms of its real *and* imaginary parts as

1, 0	0, 0	0, 0	1, 0	input row 1
0, 0	0, 0	0, 0	0, 0	input row 2
1, 0	0, 0	0, 0	1, 0	input row 3
0, 0	0, 0	0, 0	0, 0	input row 4

The corresponding transform for each row as given by FFT was

2, 0	1, 1	0, 0	1, −1	output row 1
0, 0	0, 0	0, 0	0, 0	output row 2
2, 0	1, 1	0, 0	1, −1	output row 3
0, 0	0, 0	0, 0	0, 0	output row 4

To get the final result, we now have to submit the above values in *column* order; that is, our first data input consists of the values 2,0 0,0 2,0 0,0. Our second input set consists of the values 1,1 0,0 1,1 0,0 etc.

The result given by FFT for each input set corresponds to the required transform and is given below

4, 0	0, 0	4, 0	0, 0
2, 2	0, 0	2, 2	0, 0
2, −2	0, 0	2, −2	0, 0

Notice that the final transform result has not been normalised; that is, each value is divided by N^2.

Extension of the above to the three-dimensional transform

Extension to three dimensions is trivial in terms of the methodology but prohibitive in computer time for any array of reasonable size, particularly for a microcomputer.

Since our function consists of three variables x, y, z we may consider the problem as computing the transform of a set of values $f(x, y, z)$ contained in a box with dimensions $x = N_1; y = N_2; z = N_3$. The procedure for computing the transform consists of dividing the box into a series of two-dimensional planes and determining the transform for each plane in the way described above using FFT. The required three-dimensional transform is then computed by calculating the transform of the plane transforms along each column — that is, in a direction vertical to each plane using FFT.

What is of particular importance is that the one-dimensional FFT has been used in each case to obtain the multi-dimensional transform: a result which stems from the separability of the variables.

10.7 The Inverse FFT

Although most of our discussion so far has concentrated on the forward Fourier transform, it turns out that the inverse transform can be easily determined through use of the discrete forward transform. To illustrate this, consider eqns (10.3) and (10.4) which are repeated below

$$F(u) = \frac{1}{N} \sum_{x=0}^{N-1} f(x)\exp(-2\pi iux/N) \qquad (10.23)$$

and

$$f(x) = \sum_{u=0}^{N-1} F(u)\exp(2\pi iux/N) \qquad (10.24)$$

Taking the complex conjugate of eqn (10.24) and dividing both sides by N yields

$$\frac{1}{N} f^*(x) = \frac{1}{N} \sum_{u=0}^{N-1} F^*(u)\exp(-2\pi iux/N) \qquad (10.25)$$

By comparing this result with eqn (10.23) we see that the right-side of eqn (10.25) is in the form of the forward Fourier transform. Thus, if we input $F^*(u)$ into an algorithm designed to compute the forward transform, the result will be the quantity $f^*(x)/N$. Taking the complex conjugate and multiplying by N yields the desired inverse $f(x)$.

For two-dimensional square arrays, we take the complex conjugate of

$$f(x, y) = \frac{1}{N^2} \sum_{u=0}^{N-1} \sum_{v=0}^{N-1} F(u, v)\exp[2\pi i(ux + vy)/N]$$

to give

$$f^*(x,y) = \frac{1}{N^2} \sum_{u=0}^{N-1} \sum_{v=0}^{N-1} F^*(u,v) \exp\left[-2\pi i(ux+vy)/N\right]$$

It follows, therefore, that if we input $F^*(u,v)$ into an algorithm designed to compute the forward transform, the result will be $f^*(x,y)$. By taking the complex conjugate of the result we obtain $f(x,y)$. In the case where $f(x)$ or $f(x,y)$ are real, the complex conjugate operation is unnecessary since $f(x) = f^*(x)$ and $f(x,y) = f^*(x,y)$ for real functions.

10.8 Computing the Transform of Real Arrays

This section explains how the Fourier transforms of two data arrays consisting of real values only may be computed simultaneously, thereby saving storage and processing times. The method exploits the property that the transform of a set of real numbers is conjugate symmetric about the origin, a condition sometimes referred to as Hermitean symmetric. This simply means that the sign of the imaginary part of the transform is reversed in the symmetric part. For example, the transform of the set of numbers 1, 2, 3, 0, 0, 4, 5, 6 is $(2.6, 0)$; $(0.48, 0.95)$; $(-0.88, 0)$; $(-0.23, 0.46)$; $(-0.38, 0)$; $(-0.23, -0.46)$; $(-0.88, 0)$; $(0.48, -0.95)$; $(2.6, 0)$.

The method works by inserting one array $r_1(n)$ as the real part and the other array $r_2(n)$ as the imaginary part of the complex array $Z(n)$. Thus, let $r_1(n)$ represent the first line of n real data values and $r_2(n)$ the second line. Then our complex number $z(n)$ is written in terms of its real and imaginary parts as

$$z(n) = r_1(n) + ir_2(n)$$

The Fourier transform of $z(n)$ is

$$Z(k) = \frac{1}{N} \sum_{n=0}^{N-1} [r_1(n) + ir_2(n)] \exp\left[-2\pi i(nk/N)\right] \qquad (10.26)$$

for $k = 0, 1, 2, \ldots, N-1$, which can of course be computed using the FFT algorithm.

As previously explained, the method utilises the result that for N even, the transformed array contains unique results only for $k = 0, 1, 2, \ldots (N/2) - 1$. To obtain $R_1(k)$ and $R_2(k)$, the Fourier transforms of $r_1(n)$ and $r_2(n)$, note that

$$\exp\left[2\pi in(N-k)/N\right] = \exp(-2ink/n)$$

since $\exp(2\pi in) = 1$ for any n. Hence, if $Z^*(k)$ is the complex conjugate of $Z(k)$, then

$$Z^*(N-k) = \sum_{n=0}^{N-1} [r_1(n) - ir_2(n)] \exp(-2\pi i nk/n) \qquad (10.27)$$

From eqns (10.26) and (10.27) it follows that

$$Z(k) + Z^*(N-k) = 2 \sum_{n=0}^{N-1} r_1(n)\exp(-2\pi i nk/N) = 2R_1(k) \qquad (10.28)$$

$$Z(k) - Z^*(N-k) = 2i \sum_{n=0}^{N-1} r_2(n)\exp(-2\pi i nk/N) = 2iR_2(k) \qquad (10.29)$$

Thus the two real-valued records, $r_1(n)$ and $r_2(n)$, have Fourier transforms $R_1(k)$ and $R_2(k)$ given by

$$R_1(k) = \frac{Z(k) + Z^*(N-k)}{2}$$

$$R_2(k) = \frac{Z(k) - Z^*(N-k)}{2}$$

for $k = 1, 2, \ldots, N-1$.

If we denote the real and imaginary parts of $Z(k)$ by $a(k)$ and $b(k)$ and respectively, we may re-write the above equation as

$$R_1(k) \;=\; \frac{a(k) + ib(k) + a(N-k) - ib(N-k)}{2}$$

$$\;=\; \frac{a(k) + a(N-k) + i\,[b(k) - b(N-k)]}{2}$$

Multiplication of eqn (10.29) by i and re-arranging gives

$$R_2(k) \;=\; \frac{b(k) + b(N-k) + i\,[a(N-k) - a(k)]}{2}$$

for $k = 1, 2, \ldots, N-1$. Thus, by re-arranging the real and imaginary parts of the resulting (single) transform of the input, we are able to derive the complete transform for the two arrays.

The program RFT, which is the FFT program modified in lines 335–375 to incorporate the above properties, is illustrated to compute the transform of the two real arrays 1, 2, 3, 0; 0, 4, 5, 6. Note that the results obtained by running FFT on each separate array differ from those of RFT by a scaling constant.

11 Applications and other transformations

This chapter illustrates application of the FFT to the processes of convolution and correlation. As described previously, these two processes play an important part in image processing. However, because the microprocessor is limited by its capacity to process most real images, of size generally at least 256×256 picture elements, we have restricted our illustration of these processes to one-dimensional situations. For example, the correlation process is important for aligning sequences, or for providing a measure of how closely related two given sequences are. This situation often occurs in biology where we are frequently required to match sequences made up of nucleic or amino acid components which play an important part in the biological growth process. An example is given to illustrate the application of sequence alignment on nucleic acid data using the FFT approach.

The chapter concludes with a discussion of certain other transformations which are used in the analysis of data. These are the Walsh and Hadamard transformations. The former transformation has certain similarities with the Fourier transform and these are illustrated and discussed. Where possible programs are provided, together with examples on their use.

11.1 Convolution

The convolution of two functions $f(x)$ and $g(x)$ is

$$\int_{-\infty}^{+\infty} f(u)g(x-u)\,du \tag{11.1}$$

and is denoted as $f(x)*g(x)$.

The concept of convolution is best illustrated graphically and is shown in figure 11.1. Before carrying out the integration, it is necessary to form the function $g(x-u)$. This is shown in figure 11.1 and corresponds to the operation of folding $g(u)$ about the origin to form $g(-u)$ and then displacing this function by x. Then, for any given value of x, we multiply $f(u)$ by the corresponding $g(x-u)$ and integrate the product from $-\infty$ to $+\infty$. In the example given, the function only exists over a finite range of values and so the integration is only carried out over this range.

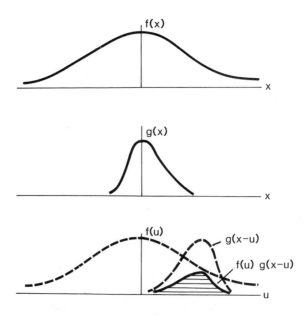

Figure 11.1. Illustration of steps in convolution process.

One useful method for illustrating the convolution process is that of graphical construction. This method relies on drawing or plotting one of the functions involved in the convolution backward on a movable piece of paper as shown in figure 11.2 and then sliding the piece of paper along in the direction of the axes of the abscissas. When the movable piece is to the left of the position shown, the product of g with f reversed is zero. By marking an arrow in some convenient position, we can keep track of this. Then suddenly, at the position shown, the integral of the product begins to assume non-zero values. By moving the paper a little further along, as indicated by a broken outline, we find that the convolution will be positive and increasing from zero approximately linearly with displacement. Further along still, we see that a maximum will occur beyond which the convolution dies away. This procedure is used in the next section to explain serial products.

Serial products

Consider two polynomials

$$a_0 + a_1x + a_2x^2 + a_3x^3 + \ldots$$

and

$$b_0 + b_1x + b_2x^2 + b_3x^3 + \ldots$$

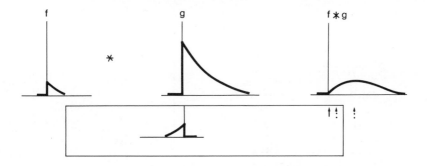

Figure 11.2. Graphical illustration of convolution process using a paper strip.

Their product is

$$a_0 b_0 + (a_0 b_1 + a_1 b_0)x + (a_0 b_2 + a_1 b_1 + a_2 b_0)x^2 + (a_0 b_3 + a_1 b_2 + a_2 b_1 + a_3 b_0)x^3 + \ldots$$

which we call

$$c_0 + c_1 x + c_2 x^2 + c_3 x^3 + \ldots$$

where

$$c_0 = a_0 b_0$$
$$c_1 = a_0 b_1 + a_1 b_0$$
$$c_2 = a_0 b_2 + a_1 b_2 + a_2 b_0$$
$$c_3 = a_0 b_3 + a_1 b_2 + a_2 b_1 + a_3 b_0$$

This elementary observation has an important connection with convolution. Suppose that two functions f and g are given and that it is required to calculate their convolution numerically. We form a sequence of values of f at short regular intervals of width w, as shown in figure 11.3. And also a corresponding sequence of values of g

$$\{g_0 g_1 g_2 g_3 \cdots g_n\}$$

We then approximate the convolution integral.

$$\int_{-\infty}^{+\infty} f(x')g(x - x') \, dx'$$

by summing products of corresponding values of f and g, taking different discrete values of x one by one. It is convenient to write the g sequence on a movable strip of paper which can be slid into successive positions relative to

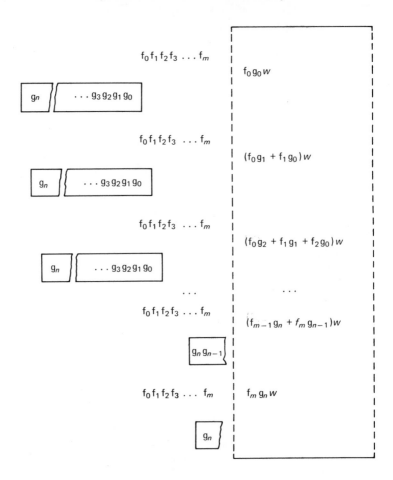

Figure 11.3

the f sequence for each successive value. The first few stages of this approach are shown in figure 11.3. It will be noticed that the g sequence has been written in reverse, as required by the formula. Since $f*g = g*f$, the f sequence could have been written in reverse, in which case *it* would have been written on the movable strip.

It will be seen that this procedure generates the same expressions that occur in the multiplication of series and we therefore introduce the term 'serial product' to describe the sequence of numbers.

$$\{f_0 g_0 \quad f_0 g_1 + f_1 g_0 \quad\quad f_0 g_2 + f_1 g_1 + f_2 g_0 \ldots\}$$

derived from the two sequences

$$\{f_0 f_1 f_2 f_3 \ldots\} \quad \text{and} \quad \{g_0 g_1 g_2 g_3 \ldots\}$$

We transfer the asterisk notation to represent this relationship between the three sequences as follows

$$\{f_0 f_1 \ldots f_m\} * \{g_0 g_1 \ldots g_n\} = \{f_0 g_0 \quad f_0 g_1 + f_1 g_0 \ldots f_m g_n\}$$

Alternatively, we may define the $(i + 1)$th term of the serial product of $\{f_i\}$ and $\{g_i\}$ to be

$$\sum_j f_j g_{i-j}$$

The importance of convolution in frequency-domain analysis lies in the fact that $f(x)*g(x)$ and $F(u)\,G(u)$ constitute a Fourier transform pair. In other words, if $f(x)$ has the Fourier transform $F(u)$ and $g(x)$ has the Fourier transform $G(u)$, then $f(x)*g(x)$ has the Fourier transform $F(u)\,G(u)$.

This result, formally stated as

$$f(x)*g(x) \Longleftrightarrow F(u)G(u) \tag{11.2}$$

indicates that convolution in the x-domain can also be obtained by taking the inverse Fourier transform of the product $F(u)G(u)$. An analogous result is that convolution in the frequency domain reduces to multiplication in the x-domain; that is

$$f(x)g(x) \Longleftrightarrow F(u)*G(u) \tag{11.3}$$

These two results are commonly referred to as the convolution theorem.

Suppose that instead of being continuous, $f(x)$ and $g(x)$ are discretised into sampled arrays of size A and B, respectively: $\{f(0), f(1), f(2), \ldots, f(A-1)\}$ and $\{g(0), g(1), g(2), \ldots, g(B-1)\}$. Now, it can be shown that the discrete Fourier transform and its inverse are periodic functions. In order to formulate a discrete convolution theorem that is consistent with this periodicity property, we may *assume* that the discrete functions $f(x)$ and $g(x)$ are periodic with some period M. The resulting convolution will then be periodic with the same period. The problem is how to select a value for M. It can be shown (Brigham, 1974) that unless we choose

$$M \geqslant A + B - 1 \tag{11.4}$$

the individual periods of the convolution will overlap; this overlap is commonly referred to as wrap-around error. If $M = A + B - 1$, the periods will be adjacent; if $M > A + B - 1$, the periods will be spaced apart, with the degree of separation being equal to the difference between M and $A + B - 1$. Since the assumed period must be greater than either A or B, the length of the sampled sequences must be increased so that both are of length M. This can be done by appending zeros to the given samples to form the following *extended* sequences

$$f_e(x) = \begin{cases} f(x) & 0 \leqslant x \leqslant A - 1 \\ 0 & A \leqslant x \leqslant M - 1 \end{cases}$$

and

$$g_e(x) = \begin{cases} g(x) & 0 \leqslant x \leqslant B - 1 \\ 0 & B \leqslant x \leqslant M - 1 \end{cases}$$

Based on this, we define the discrete convolution of $f_e(x)$ and $g_e(x)$ by the expression

$$f_e(x) * g_e(x) = \sum_{m=0}^{M-1} f_e(m)\, g_e(x - m) \tag{11.5}$$

for $x = 0, 1, 2, \ldots, m - 1$. The convolution function is a discrete, periodic array of length M, with the values $x = 0, 1, 2, \ldots, M - 1$ describing a full period of $f_e(x) * g_e(x)$.

The mechanics of discrete convolution are basically the same as for continuous convolution. The only differences are that displacements take place in discrete increments corresponding to the separation between samples, and that integration is replaced by a summation. Similarly, eqns (11.2) and (11.3) also hold in the discrete case where, to avoid wrap-around error, we use $f_e(x)$ and its transform. The discrete variables x and u assume values in the range $0, 1, 2, \ldots, M - 1$.

Two discrete functions $f_e(m)$ and $g_e(m)$, shown in figure 11.4, are used to illustrate the procedure. The figure shows A samples for both $f(x)$ and $g(x)$ in the interval $[0, 1]$ as well as an assumed period of $M = A + B - 1 = 2A - 1$.

It is noted that the convolution function is periodic and that, since $M = 2A - 1$, the periods are adjacent. Choosing $M > 2A - 1$ would have produced a larger separation between these periods. It is also important to note that a period is completely described by M samples. Further note that the product corresponding to the transform values $F(u), G(u)$ is expressed by the multiplication of two complex numbers. Thus if a, b and c, d correspond to the real and imaginary parts of $F(u)$ and $G(u)$ respectively, the complex number formed by the product $F(u)\, G(u)$ has real and imaginary parts: $ac-bd; ad-bc$.

The program CONVFT is provided to illustrate the determination of convolution of two discrete functions. The values used as input to the program are determined by the functions defined as

$$f_e(m) \begin{cases} 1 & m = 0, \ldots, 3 \\ 0 & m = 4, \ldots, 8 \end{cases}$$

$$g_e(m) \begin{cases} 0.5 & m = 0, \ldots, 3 \\ 0 & m = 4, \ldots, 8 \end{cases}$$

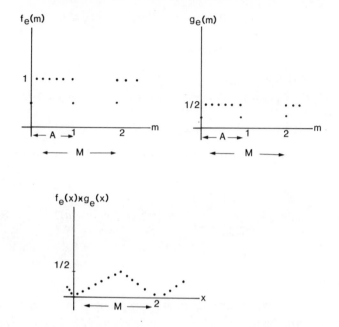

Figure 11.4. Illustration of convolution for two discrete functions.

Two-dimensional convolution is analogous in form to eqn (11.1). Thus, for two functions $f(x,y)$ and $g(x,y)$, we have

$$f(x,y)*g(x,y) \int_{-\infty}^{+\infty}\!\!\int f(\alpha,\beta)g(x-\alpha,y-\beta)\,d\alpha\,d\beta \tag{11.6}$$

The convolution theorem in two-dimensions then is given by the relations

$$f(x,y)*g(x,y) \iff F(u,v)G(u,v) \tag{11.7}$$

and

$$f(x,y)g(x,y) \iff F(u,v)*G(u,v) \tag{11.8}$$

The two-dimensional, discrete convolution is formulated by letting $f(x,y)$ and $g(x,y)$ be discrete arrays of size $A \times B$ and $C \times D$ respectively. As in the one-dimensional case, these arrays must be assumed periodic with some period M and N in the x and y directions respectively. Wrap-around error in the individual convolution periods is avoided by choosing:

$$M \geqslant A + C - 1 \tag{11.9}$$

and

$$N \geqslant B + D - 1 \tag{11.10}$$

The periodic sequences are formed by extending $f(x,y)$ and $g(x,y)$ as follows

$$f_e(x,y) = \begin{cases} f(x,y) & 0 \leqslant x \leqslant A-1 \quad \text{and} \quad 0 \leqslant y \leqslant B-1 \\ 0 & A \leqslant x \leqslant M-1 \quad \text{or} \quad B \leqslant y \leqslant N-1 \end{cases}$$

and

$$g_e(x,y) = \begin{cases} g(x,y) & 0 \leqslant x \leqslant C-1 \quad \text{and} \quad 0 \leqslant y \leqslant D-1 \\ 0 & C \leqslant x \leqslant M-1 \quad \text{or} \quad D \leqslant y \leqslant N-1 \end{cases}$$

The two-dimensional convolution of $f_e(x,y)$ and $g_e(x,y)$ is given by the relation

$$f_e(x,y) * g_e(x,y) = \sum_{m=0}^{M-1} \sum_{n=0}^{N-1} f_e(m,n)\, g_e(x-m, y-n) \tag{11.11}$$

for $x = 0, 1, 2, \ldots, M-1$

and $y = 0, 1, 2, \ldots, N-1$

The $M \times N$ array given by this equation is one period of the discrete, two-dimensional convolution. If M and N are chosen according to eqns (11.9) and (11.10), this array is guaranteed to be free of interference from other adjacent periods. As in the one-dimensional case, the continuous convolution theorem given in eqns (11.7) and (11.8) also applies to the discrete case with $u = 0, 1, 2, \ldots, M-1$ and $v = 0, 1, 2, \ldots, N-1$. All computations involve the extended functions $f_e(x,y)$ and $g_e(x,y)$.

11.2 Correlation

The correlation of two continuous functions $f(x)$ and $g(x)$ denoted by $f(x) \circ g(x)$, is defined by the relation

$$f(x) \circ g(x) = \int_{-\infty}^{\infty} f(\alpha) g(x+\alpha)\, d\alpha \tag{11.12}$$

The forms of eqns (11.12) and (11.1) are similar, the only difference being that the function $g(x)$ is not folded about the origin. Thus, to perform correlation we simply slide $g(x)$ by $f(x)$ and integrate the product from $-\infty$ to $+\infty$ for each value of displacement x.

The discrete equivalent of eqn (11.12) is defined as

$$f_e(x) \circ g_e(x) = \sum_{m=0}^{M-1} f_e(m)\, g_e(x+m) \tag{11.13}$$

for $x = 0, 1, 2, \ldots, M - 1$

The same comments made above with regard to $f_e(x)$ and $g_e(x)$, the assumed periodicity of these functions and the choice of values for M, also apply to eqn (11.13).

It is important to note that when the two functions $f(x)$ and $g(x)$ are the same, eqn (11.12) is called the auto correlation function; if $f(x)$ and $g(x)$ are different, the term cross-correlation is normally used.

Similar expressions to the above equations hold for two dimensions. Thus, if $f(x, y)$ and $g(x, y)$ are functions of continuous variables, their correlation is defined as

$$f(x, y) \circ g(x, y) = \int_{-\infty}^{\infty} \int_{-\infty}^{\infty} f(\alpha, \beta) g(x + \alpha, y + \beta) \, d\alpha \, d\beta \qquad (11.14)$$

and for the discrete case

$$f_e(x, y) \circ g_e(x, y) = \sum_{m=0}^{M-1} \sum_{n=0}^{N-1} f_e(m, n) g_e(x + m, y + n) \qquad (11.15)$$

for $x = 0, 1, 2, \ldots, M - 1$ and $y = 0, 1, 2, \ldots, N - 1$.

As in the case of discrete convolution, $f_e(x, y)$ and $g_e(x, y)$ are extended functions, and M and N are chosen according to eqns (11.9) and (11.10) in order to avoid wrap-around error in the periods of the correlation function.

It can be shown for both the continuous and discrete cases that the following correlation theorem holds

$$f(x, y) \circ g(x, y) \iff F(u, v) \, G^*(u, v) \qquad (11.16)$$

and

$$f(x, y) \, g^*(x, y) \iff F(u, v) \circ G(u, v) \qquad (11.17)$$

where '*' represents the complex conjugate. It is understood that, when interpreted for discrete variables, all functions are assumed to be extended and periodic.

As in the case of discrete convolution, the computation of $f_e(x, y) \circ g_e(x, y)$ is often more efficiently carried out in the frequency domain using the FFT algorithm to obtain the forward and inverse transforms. The merits of using the FFT approach as opposed to the direct application of either of eqns (11.13) or (11.15) relies on the fact that in one case we are dealing with a number of operations proportional to N^2 (the direct approach as opposed to $N \log_2 N$ using the FFT). Specifically, the time required to compute eqn (11.13) using the FFT approach is equal to $2c_1 N \log_2 N$ (that is, the number of operations required to evaluate the transforms of $f_e(m)$ and $g_e(m)$) plus the number of operations '$c_2 N$' to perform the multiplication of the respective transforms (where c_1 and

c_2 are constants dependent upon the particular computer being used) together with the number of operations $c_1 N \log_2 N$ required for the inverse transform, thus giving a total of $C = 3c_1 N \log_2 N + c_2 N$ operations. (N is of course assumed to be a power of 2.)

Table 11.1 gives approximate values for the computational advantage in determining convolution/correlation by use of the FFT opposed to direct calculation. (The values for c_1 and c_2 in this calculation have been assumed equal to unity.) Since the results show that the FFT approach is significantly faster for values of N greater than 32, a result independently determined by Brigham (1974), this approximation is not considered critical in illustrating the relative computational advantage to be gained through use of the FFT.

Table 11.1

N	N^2	C	Advantage (N^2/c)
2	4	8	0.5
4	16	28	0.57
8	64	80	0.80
16	256	208	1.23
32	1024	512	2.00
64	4096	1216	3.37
128	16384	2816	5.82
256	65536	6400	10.24
512	262144	14336	18.29
1024	1048576	31744	33.03
2048	4194304	69632	60.24
4096	16777216	151552	110.70
8192	67108864	327680	204.80

To illustrate the application of the correlation process in a practical context we have chosen an example from the biological sciences in which image processing and analysis play an increasingly important part in helping scientists to understand the complex chemical structures present in living cells.

Many readers will already be aware that most of the information essential to cellular growth and development is contained in chemical structures whose components are made up of nucleic acids. For example, the well-known structure DNA consists of nucleic acids arranged in a helical pattern. Much of the current work therefore aimed at understanding the different mechanisms of cellular behaviour requires details of the differences between chemical structures in terms of their nucleic acid composition. As we have already shown, the matching or correlation between two sequences or data sets can be achieved using the FFT approach described in section 11.1. The following example, selected from the work of Felsenstein *et al.* (1982), who first suggested the approach, illustrates how the correlation procedure can be adapted to investigate the matching of different nucleic acid sequences.

Structures consisting of nucleic acid sequences are usually represented by a series of letters in which each letter corresponds to the name of the particular nucleic acid (commonly called a base) which occurs at that position in the sequence. Part of a typical sequence appears as AACGUGGC, where A stands for the nucleic acid bases adenosine, C for cytosine, G for guanine and U for uracil. (In practice, sequences may consist of several thousand bases.)

The procedure involves reading each sequence and then constructing a series of indicators, one for each of the four bases. Each of these is an array containing a one whenever that base exists in the corresponding sequence and a zero where it does not exist. For example, the above sequence has the four indicator sequence

Sequence	A	A	C	G	U	G	G	C
Indicator for A	1	1	0	0	0	0	0	0
Indicator for C	0	0	1	0	0	0	0	1
Indicator for G	0	0	0	1	0	1	1	0
Indicator for U	0	0	0	0	1	0	0	0

We denote the jth entry in the indicator function for A as $X_j^{(A)}$ and similarly for the other bases we have $X_j^{(C)}, X_j^{(G)}, X_j^{(U)}$. The corresponding indicator functions for the sequence available for comparison will be $Y_j^{(A)}, Y_j^{(C)}, Y_j^{(G)}, Y_j^{(U)}$.

The number of matches of A's when the second sequence is displaced by k from the first is then given by

$$Z_k^{(A)} = \sum_{j=1}^{n} X_j^{(A)} Y_{j+k}^{(A)}$$

(Compare this with eqn (11.13) the standard correlation equation.) The overall number of matches at a shift of k is given by

$$Z_k = Z_k^{(A)} + Z_k^{(C)} + Z_k^{(G)} + Z_k^{(U)}$$

Thus the procedure for evaluating Z_k involves the following steps. First compute the Fourier transforms of the indicator functions of the two sequences. This will be eight Fourier transforms in all. Since from the theory outlined earlier in this section we know that if X and Y are sequences whose discrete Fourier transforms are U and V, then the sequence Z giving the number of matches has the Fourier transfor W, where

$$W_j = U_j V_j^*$$

we could now use the above equation to determine the Fourier transforms

$W_j^{(A)}, \ldots, W_j^{(U)}$, which are the transforms of the number of matches of A's, C's, G's and U's at all possible shifts.

If our interest is in computing the overall number of matches, we can sum the four W's to get

$$W_j = W_j^{(A)} + W_j^{(C)} + W_j^{(G)} + W_j^{(U)}$$

We now take the inverse Fourier transforms of the sequence W_j. The real parts of the resulting sequence of complex numbers will be the numbers of matches at shifts $0, 1, \ldots, n$. We can thus obtain the desired result for each base in a total of nine Fourier transforms. We must of course bear in mind the practical limitations imposed in this procedure using the FFT algorithm which is available — namely, the length of the sequence must be a power of 2 (obtained by adding zeros to the data up to the next power of 2) and the doubling of the length of the sequences with zeros, as explained previously, to avoid wrap-around.

11.3 Other Transformations

Of the various transformations which currently exist and are, in fact, used in image and data analysis, two which are of particular interest are the Walsh and Hadamard transforms. The Walsh transform is interesting because it shares certain similarities with the Fourier transform and can be programmed for the microcomputer using the same principles (successive doublings) as used for the FFT. The Hadamard transform is similar in some respects to the Walsh transform but since the successive doubling principle cannot be utilised, its general applicability to data analysis via the microcomputer is strictly limited. However, the transformation together with the appropriate algorithm have been included here to enable the reader to at least gain familiarity with its use on data sets of restricted size.

Before proceeding with the general formulation of the above transformations, an example is given of the Walsh transform to illustrate its properties and similarities with the Fourier transform. Figure 11.5 shows the respective input data, FFT power spectrum and FWT (fast Walsh transform) power spectrum for three types of data. In (a) the input data consists of a simple cosine curve, whereas in (b) and (c) the data consists of a set of regular pulses distributed to lie along a cosine curve. Comparison between the FFT and FWT spectra show a certain degree of coincidence between them in relation to the relative position of certain of the peaks. However, it can be seen that the Walsh transform gives not only peaks which are not as detailed as those determined by the Fourier transform but also gives additional peaks not representative of the particular spatial frequency characteristics of the original input data. In fact, the findings by several workers in the field of image and data analysis confirm that, in general, the Walsh power spectra are more diffuse, less peaked and generally less detailed than the Fourier spectra. However, since the FWT is substantially faster than the FFT it offers an attractive method for data analysis in the case of very large data sets. Since its correlation with the FFT is only partial, however, care must be exercised when interpreting results, particularly those concerned with frequency evaluation.

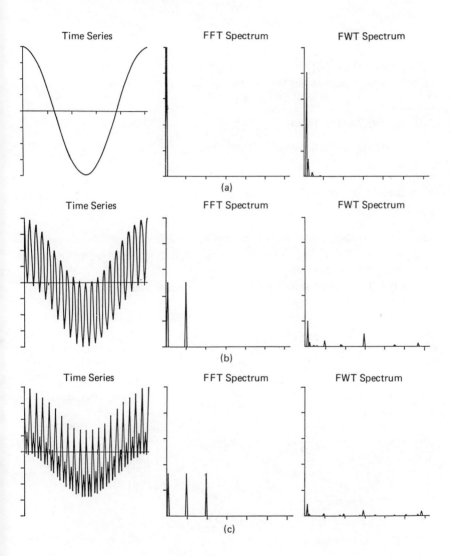

Figure 11.5. Comparison between Fourier and Walsh spectra.

General formulation

The one-dimensional discrete Fourier transform is one of a class of important transforms which can be expressed in terms of the general relation

$$T(u) = \sum_{x=0}^{N-1} f(x)g(x,u) \qquad (11.18)$$

where $T(u)$ is the transform of $f(x)$, $g(x,u)$ is the forward transformation kernel and u assumes values in the range $0, 1, \ldots, N-1$. In the case of the discrete Fourier transform, for example

$$g(x,u) = \exp(-2\pi iux/N)$$

The inverse transform is given by the relation

$$f(x) = \sum_{u=0}^{N-1} T(u)h(x,u) \qquad (11.19)$$

where $h(x,u)$ is the inverse transformation kernel and x assumes values in the range $0, 1, \ldots, N-1$. The nature of a transform is determined by the properties of its transformation kernel.

For two-dimensional square arrays the forward and inverse transforms are given by the equations

$$T(u,v) = \sum_{x=0}^{N-1} \sum_{y=0}^{N-1} f(x,y)g(x,y,u,v) \qquad (11.20)$$

and

$$f(x,y) = \sum_{u=0}^{N-1} \sum_{v=0}^{N-1} T(u,v)h(x,y,u,v) \qquad (11.21)$$

where, as above, $g(x,y,u,v)$ and $h(x,y,u,v)$ are called the forward and inverse transformation kernels, respectively.

The forward kernel is said to be separable if

$$g(x,y,u,v) = g_1(x,u)g_2(y,v)$$

The kernel is, in addition, symmetric if g_1 is functionally equal to g_2. In this case, the above equation can be expressed in the form

$$g(x,y,u,v) = g_1(x,u)g_1(y,v)$$

Identical comments hold for the inverse kernel if $g(x,y,u,v)$ is replaced by $h(x,y,u,v)$ in the above equations.

The two-dimensional Fourier transform is a special case of eqn (11.20). It has the kernel

$$g(x,y,u,v) = 1/N \exp\left[-2\pi i(ux + vy)/N\right]$$

This is both separable and symmetric since letting $ux/N = \theta_1$ and $vy/N = \theta_2$ we have

$$
\begin{aligned}
g(x,y,u,v) &= 1/N \exp\left[-2\pi i(\theta_1 + \theta_2)\right] \\
&= 1/N\left[\cos 2\pi(\theta_1 + \theta_2) - i\sin 2\pi(\theta_1 + \theta_2)\right] \\
&= 1/N\left[\cos 2\pi\theta_1 \cos 2\pi\theta_2 - \sin 2\pi\theta_1 \sin 2\pi\theta_2\right. \\
&\quad \left. - i(\sin 2\pi\theta_1 \cos 2\pi\theta_2 + \cos\theta_1 + \cos 2\pi\theta_1 \sin 2\pi\theta_2)\right] \\
&= 1/\sqrt{(N)}\left[\cos 2\pi\theta_1 - i\sin 2\pi\theta_1\right]\, 1/\sqrt{(N)}\left[\cos 2\pi\theta_2 - i\sin 2\pi\theta_2\right] \\
&= 1/\sqrt{(N)}\left[\exp-2\pi iux/N\right]\, 1/\sqrt{(N)}\left[\exp(-2\pi ivy/N)\right]
\end{aligned}
$$

It can be similarly shown that the inverse Fourier kernel is also separable and symmetric.

A transform with a separable kernel can be computed in two steps, each requiring a one-dimensional transform. First, the one-dimensional transform is taken along each row of $f(x,y)$, yielding

$$T(x,v) = \sum_{y=0}^{N-1} f(x,y)g_2(y,v) \tag{11.22}$$

for $x, v = 0, 1, 2, \ldots, N-1$. Next, the one-dimensional transform is taken along each column of $T(x,v)$; this results in the expression

$$T(u,v) = \sum_{x=0}^{N-1} T(x,v)g_1(x,u) \tag{11.23}$$

for $u, v = 0, 1, 2, \ldots, N-1$.

This procedure agrees with the approach given in chapter 10 for the Fourier transform. The same final results are obtained if the transform is taken first along each column of $f(x,y)$ to obtain $T(y,u)$ and then along each row of the latter function to obtain $T(u,v)$.

11.4 Walsh Transform

When $N = 2^n$, the discrete Walsh transform of a function f(x), denoted by W(u), is obtained by substituting the kernel

$$g(x, u) = 1/N \prod_{i=0}^{n-1} (-1)^{b_i(x) b_{n-1-i}(u)} \qquad (11.24)$$

into eqn 11.18 — that is

$$W(u) = 1/N \sum_{x=0}^{n-1} f(x) \prod_{i=0}^{n-1} (-1)^{b_i(x) b_{n-1-i}(u)} \qquad (11.25)$$

where $b_k(Z)$ is the kth bit in the binary representation of Z. For example, if $n = 3$ and $Z = 2$ (010 in binary), we have that $b_0(2) = 0, b_1(2) = 1, b_2(2) = 0$. Thus for $x = 2$ and $u = 2$ (with $n = 3$) we have

$$g(x, u) = g(2, 2) = 1/N \prod_{i=0}^{2} (-1)^{b_i(2) b_{2-i}(2)}$$

$$= 1/N \{ (-1)^{b_0(2) b_2(2)} (-1)^{b_1(2) b_1(2)} (-1)^{b_2(2) b_0(2)} \}$$

$$= 1/N \{ (-1)^{0 \times 0} (-1)^{1 \times 1} (-1)^{0 \times 0} \}$$

$$= 1/N (-1)$$

Similarly for $x = 3$ and $u = 6$ (with $n = 3$) we have

$$b_0(3) = 1, b_1(3) = 1, b_2(3) = 0$$

$$b_0(6) = 0, b_1(6) = 1, b_2(6) = 1$$

since the binary representation of 3 and 6 is 011 and 110 respectively.

$$g(x, u) = g(3, 6) = 1/N \prod_{i=0}^{2} (-1)^{b_i(3) b_{2-i}(6)}$$

$$= 1/N \{ (-1)^{b_0(3) b_2(6)} (-1)^{b_1(3) b_1(6)} (-1)^{b_2(3) b_0(6)} \}$$

$$= 1/N \{ (-1)^{1 \times 1} (-1)^{1 \times 1} (-1)^{0 \times 0} \}$$

$$= 1/N \{ + 1 \}$$

The program WALSH determines the WALSH transform for a given function f(x) for $N = 8$. The program is illustrated in relation to the function f(x) defined in chapter 10 (section 10.1) and the values of $g(x, u)$ excluding the $1/N$ constant term are shown in table 11.2. These values were determined from the program WALSH by printing the matrix of values in $G(X, U)$.

Table 11.2

	0	1	2	3	4	5	6	7
0	+	+	+	+	+	+	+	+
1	+	+	+	+	−	−	−	−
2	+	+	−	−	+	+	−	−
3	+	+	−	−	−	−	+	+
4	+	−	+	−	+	−	+	−
5	+	−	+	−	−	+	−	+
6	+	−	−	+	+	−	−	+
7	+	−	−	+	−	+	+	−

The inverse kernel is identical to the forward kernel by virtue of the orthogonal properties of the rows and columns in the kernel array except for the constant multiplicative factor of $1/N$; that is

$$h(x, u) = \prod_{i=0}^{n-1} (-1)^{b_i(x)b_{n-1-i}(u)} \tag{11.26}$$

Thus the inverse Walsh transform is given by

$$f(x) = \sum_{u=0}^{N-1} W(u) \prod_{i=0}^{n-1} (-1)^{b_i(x)b_{n-1-i}(u)} \tag{11.27}$$

From eqns (11.25) and (11.27) above it can be seen that the forward and inverse Walsh transforms differ only by the $1/N$ term. Thus, any algorithm for computing the forward transform can be used directly to obtain the inverse transform simply by multiplying the result of the algorithm by N.

The two-dimensional forward and inverse Walsh kernels are given by the relations

$$g(x, y, u, v) = 1/N \prod_{i=0}^{n-1} (-1)^{[b_i(x)b_{n-1-i}(u) + b_i(y)b_{n-1-i}(v)]} \tag{11.28}$$

$$h(x, y, u, v) = 1/N \prod_{i=0}^{n-1} (-1)^{[b_i(x)b_{n-1-i}(u) + b_i(y)b_{n-1-i}(v)]} \tag{11.29}$$

Since the formulation given in these equations yields identical kernels, it follows from eqns (11.20) and (11.21) that the forward and inverse Walsh transforms are also equal in form; that is

$$W(u,v) = 1/N \sum_{x=0}^{N-1} \sum_{y=0}^{N-1} f(x,y) \prod_{i=0}^{n-1} (-1)^{[b_i(x)b_{n-1-i}(u) + b_i(y)b_{n-1-i}(v)]}$$

(11.30)

and

$$f(x,y) = 1/N \sum_{u=0}^{N-1} \sum_{v=0}^{N-1} W(u,v) \prod_{i=0}^{n-1} (-1)^{[b_i(x)b_{n-1-i}(u) + b_i(y)b_{n-1-i}(v)]}$$

(11.31)

Thus any algorithm which is used to compute the two-dimensional forward Walsh transform can also be used without modification to compute the inverse transform.

The Walsh transform kernels are separable and symmetric since

$$g(x,y,u,v) = g_1(x,u)g_1(y,v)$$

$$= h_1(x,u)h_1(y,v)$$

$$g(x,y,u,v) = \left[\frac{1}{\sqrt{N}} \prod_{i=0}^{n-1} (-1)^{b_i(x)b_{n-1-i}(u)} \right] \left[\frac{1}{\sqrt{N}} \prod_{i=0}^{n-1} (-1)^{b_i(y)b_{n-1-i}(v)} \right]$$

(11.32)

It follows, therefore, that $W(u,v)$ and its inverse can be computed by successive applications of the one-dimensional Walsh transform given in eqn (11.25). The procedure followed in the computation is the same as the one given in chapter 10 (section 10.6) for the Fourier transform.

The Walsh transform can be computed by a fast algorithm identical in form to the successive-doubling method given in chapter 10 for the FFT. The only difference is that all exponential terms W_N are set equal to one in the case of the fast Walsh transform. Equations (10.13) and (10.14) which are the basic relations leading to the FFT then become:

$$W(u) = \tfrac{1}{2}\{W_{even}(u) + W_{odd}(u)\}$$

(11.33)

$$W(u + M) = \tfrac{1}{2}\{W_{even}(u) - W_{odd}(u)\}$$

(11.34)

where $M = N/2, u = 0, 1, \ldots, M - 1$, and $W(u)$ denotes the one-dimensional Walsh transform.

To illustrate the above, consider the case of the Walsh transform given by eqn (11.30) for $N = 4$.

$$W(0) = \frac{1}{4} \sum_{x=0}^{3} [f(x) \prod_{i=0}^{1} (-1)^{b_i(x)b_{1-i}(0)}]$$

$$= \frac{1}{4} [f(0) + f(1) + f(2) + f(3)]$$

$$W(1) = \frac{1}{4} \sum_{x=0}^{3} [f(x) \prod_{i=0}^{1} (-1)^{b_i(x)b_{1-i}(1)}]$$

$$= \frac{1}{4} [f(0) - f(1) - f(2) - f(3)]$$

$$W(2) = \frac{1}{4} \sum_{x=0}^{3} [f(x) \prod_{i=0}^{1} (-1)^{b_i(x)b_{1-i}(2)}]$$

$$= \frac{1}{4} [f(0) - f(1) + f(2) - f(3)]$$

$$W(3) = \frac{1}{4} \sum_{x=0}^{3} [f(x) \prod_{i=0}^{1} (-1)^{b_i(x)b_{1-i}(3)}]$$

$$= \frac{1}{4} [f(0) - f(1) - f(2) + f(3)]$$

In order to show the validity of eqns (11.33) and (11.34) we subdivide these results into two groups; that is

$$W_{even}(0) = \tfrac{1}{2}[f(0) + f(2)] \quad W_{odd}(0) = \tfrac{1}{2}[f(1) + f(3)]$$
$$W_{even}(1) = \tfrac{1}{2}[f(0) - f(2)] \quad W_{odd}(1) = \tfrac{1}{2}[f(1) - f(3)]$$

From eqn (11.33) we have

$$W(0) = \tfrac{1}{2}[W_{even}(0) + W_{odd}(0)]$$
$$= \tfrac{1}{4}[f(0) + f(1) + f(2) + f(3)]$$

and

$$W(1) = \tfrac{1}{2}[W_{even}(1) + W_{odd}(1)]$$
$$= \tfrac{1}{4}[f(0) + f(1) - f(2) - f(3)]$$

The next two terms are computed from these results using eqn (11.34)

$$W(2) = \tfrac{1}{2}[W_{even}(0) - W_{odd}(0)]$$
$$= \tfrac{1}{4}[f(0) - f(1) + f(2) - f(3)]$$

128 *Spatial Structure and the Microcomputer*

$$W(3) = \tfrac{1}{2}[W_{even}(1) - W_{odd}(1)]$$

$$= \tfrac{1}{4}[f(0) - f(1) - f(2) + f(3)]$$

Thus, computation of $W(u)$ by eqn (11.25) or by eqns (11.33) and (11.34) yields identical results.

As indicated above, an algorithm used to compute the FFT by the successive-doubling method can be easily modified for computing a fast Walsh transform simply by setting all trigonometric terms equal to one.

11.5 Hadamard Transform

When $N = 2^n$ the one-dimensional forward Hadamard kernel is given by the relation

$$g(x, u) = 1/N(-1)^{\sum_{i=0}^{n-1} b_i(x)b_i(u)} \tag{11.35}$$

where as in the Walsh transform $b_k(Z)$ is the kth bit in the binary representation of Z. Thus if $n = 3$ and $Z = 2$ (010 in binary), we have that $b_0(2) = 0$, $b_1(2) = 1, b_2(2) = 0$. Thus for $x = 2$ and $u = 2$ (with $n = 3$) we have

$$g(x, u) = g(2, 2) = 1/N(-1)^{\sum_{i=0}^{2} b_i(2)b_i(2)}$$

$$= 1/N\{(-1)^{(00 + 11 + 00)}\}$$

$$= 1/N$$

Substitution of eqn (11.35) into eqn (11.18) yields the following expression for the one-dimensional Hadamard transform

$$H(u) = 1/N \sum_{x=0}^{N-1} f(x)(-1)^{\sum_{i=0}^{n-1} b_i(x)b_i(u)} \tag{11.36}$$

where $N = 2^n$ and u assumes values in the range $0, 1, 2, \ldots, n - 1$.

As in the case of the Walsh transform, the Hadamard kernel forms a matrix whose rows (and columns) are orthogonal. This condition again leads to an inverse kernel which, except for the $1/N$ term, is equal to the forward Hadamard kernel, that is

$$h(x, u) = (-1)^{\sum_{i=0}^{n-1} b_i(x)b_i(u)}$$

Substitution of this kernel into eqn (11.19) yields the following expression for the inverse Hadamard transform

$$f(x) = \sum_{u=0}^{N-1} H(u)(-1)^{\sum_{i=0}^{n-1} b_i(x)b_i(u)} \tag{11.37}$$

$$x = 0, 1, \ldots, N-1$$

The two-dimensional kernals are similarly given by the relations

$$g(x,y,u,v) = 1/N(-1)^{\sum_{i=0}^{n-1}[b_i(x)b_i(u)+b_i(y)b_i(v)]} \tag{11.38}$$

and

$$h(x,y,u,v) = 1/N(-1)^{\sum_{i=0}^{n-1}[b_i(x)b_i(u)+b_i(y)b_i(v)]} \tag{11.39}$$

Note that, as in the case of the Walsh transform, the two-dimensional Hadamard kernels are identical.

Substitution of eqns (11.38) and (11.39) into eqns (11.20) and (11.21) yields the following two-dimensional Hadamard transform pair

$$H(u,v) = 1/N \sum_{x=0}^{N-1} \sum_{y=0}^{N-1} f(x,y)(-1)^{\sum_{i=0}^{n-1}[b_i(x)b_i(u)+b_i(y)b_i(v)]}$$

and

$$f(x,y) = 1/N \sum_{u=0}^{N-1} \sum_{v=0}^{N-1} H(u,v)(-1)^{\sum_{i=0}^{n-1}[b_i(x)b_i(u)+b_i(y)b_i(v)]}$$

Since the forward and inverse transforms are identical, an algorithm used for computing $H(u,v)$ can be used without modification to obtain $f(x,y)$ and vice versa.

The program HADAMARD is provided to evaluate the linear function $H(u)$ for $N = 8$ and is illustrated in relation to calculating the transform of the function defined in chapter 10 (section 10.1).

The chapter concludes with a discussion on the program IMAGES which determines the Fourier transform of a variety of shapes and displays the result as an amplitude matrix. The program has been included in this section of the book since it can be easily adapted to calculate the convolution, correlation or Walsh transforms on a variety of shapes using the methods discussed previously.

The program functions by filling certain elements of a 16×16 matrix in a way which approximates to the required shape. The FFT of this matrix is then

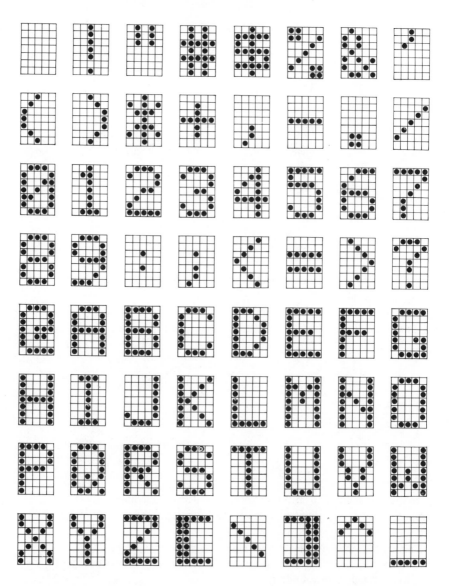

Figure 11.6. Examples showing construction of letters and digits within a two-dimensional matrix.

calculated and the two-dimensional Fourier amplitude spectrum is calculated at each point of the resulting transform matrix. This matrix is then scaled so that its values lie between zero and nine and are subsequently printed.

The main disadvantage with the program as it stands is that, by scaling the output values between zero and nine, the numerical accuracy of the result is reduced. This can of course be rectified by simply omitting the scaling procedure or setting the scale factor SC equal to unity. The output, although more accurate, is less acceptable from a graphical standpoint.

The advantage of the program is that the user is able to build up a series of different images by simply setting values for the matrix elements in lines 105–140 of the main program.

The program is demonstrated on two images. The first is a representation of the letter T and the second a simple rectangle. The number '5' has been used to fill in the appropriate elements of the matrix and so emulate the shape. Clearly, other numbers could equally well be used and the non-sensitivity of the transform spectrum to interchanging '5' with '1' is demonstrated. Figure 11.6 shows a list of characters and their matrix representation which readers will find useful as a guide to constructing different shapes.

Appendix: Computer programs for numerical and geometrical analysis

The following programs, together with documentation, are provided to illustrate many of the methods discussed in the text. Simple examples are provided to demonstrate the running of programs. No attempt to format the input/output has been made, as this can be easily undertaken by the reader for the particular microcomputer being used.

The programs are written in BASIC (except for certain of those given in the section on APL) and, although working on specific computers, can readily be adapted for the variants of BASIC available on specific microcomputers. We have used mostly the Microsoft BASIC on a NASCOM II microcomputer and the BASIC available on a BBC micro. In almost all cases specimen printout is given so that implementation on other machines may be checked.

BINARY

It is sometimes necessary to convert a decimal integer to a binary integer — a sequence of 0s and 1s. In an 8-bit BASIC only integers up to 32 768 may be represented.

```
10 REM program name "BINARY"
20 REM to convert a decimal number to binary
30 REM number should be less than 2^NI
40 REM if NI=15 number should be less than 32768
50 REM note use of INTEGER DIVISION "\"
60 DEFINT I-N
70 REM number of binary digits =NI
80 NI=15
90 DIM IND(NI)
100 PRINT "input number"
110 INPUT N
120 PRINT N,
130 FOR I=NI TO 1 STEP -1
140 I2=2^(I-1)
150 IND(I)=0
160 IF N\I2>0 THEN N=N-I2: IND(I)=1
170 PRINT IND(I);
180 NEXT I
```

CHOLESKI

It is possible to factorise a symmetric, positive definite matrix into the product
of a lower triangular matrix and its transpose. The program does this. Positive
definite means that all determinants which may be made by striking out rows
and columns of the matrix are non-negative.

```
100 REMark program name CHOLESKI
110 REMark factorisation of a symmetric
120 REMark positive definite matrix A
130 REMark into the product of a lower
140 REMark triangular matrix L and its transpose.
150 REMark L may overwrite A if required.
160 REMark if any term L(i,i) is imaginary,
170 REMark then matrix is not quite pos. def.
180 n=5
190 DIM a(n,n),l(n,n)
200 RESTORE
210 FOR i=1 TO n
220 FOR j=i TO n
230 READ a(i,j)
240 a(i,j)=a(i,j)*a(i,j)
250 a(j,i)=a(i,j)
260 l(i,j)=0
270 l(j,i)=0
280 NEXT j
290 NEXT i
300 REMark subtract borders
310 FOR i= 2 TO n
320 FOR j=1 TO n
330 a(i,j)=a(i,j)-a(1,j)
340 NEXT j
350 NEXT i
360 FOR j=2 TO n
370 FOR i=1 TO n
380 a(i,j)=a(i,j)-a(i,1)
390 NEXT i
400 NEXT j
410 REMark reduce dimensions by 1
420 n=n-1
430 FOR i= 1 TO n
440 FOR j=1 TO n
450 a(i,j)=-.5*a(i+1,j+1)
460 NEXT j
470 NEXT i
480 PRINT £5, "MATRIX TO BE FACTORISED"
490 FOR i=1 TO n
500 FOR j=1 TO n
510 PRINT £5, a(i,j),
520 NEXT j
530 PRINT £5
540 NEXT i
550 PRINT £5
560 REMark program proper begins
570 l(1,1)=SQRT(a(1,1))
580 FOR i=2 TO n
590 FOR j=1 TO i
600 IF i=j THEN GO TO 670
610 p=0
620 FOR k=1 TO i-1
630 p=p+l(i,k)*l(j,k)
640 NEXT k
650 l(i,j)=(a(i,j)-p)/l(j,j)
```

```
660 GO TO 730
670 p=0
680 FOR k=1 TO i-1
690 p=p+l(i,k)*l(i,k)
700 NEXT k
710 l(i,i)=SQRT(ABS(a(i,i)-p))
720 IF a(i,i)-p <0 THEN PRINT "L(";i;")";" imaginary"
730 NEXT j
740 NEXT i
750 REMark program proper ends
760 FOR i=1 TO n
770 PRINT £5,i;"        ",
780 FOR j=1 TO n
790 PRINT £5,l(i,j);"   ";
800 NEXT j
810 PRINT £5
820 NEXT i
830 REMark find determinant
840 det = 1
850 FOR i=1 TO n
860 det=det*l(i,i)
870 NEXT i
880 PRINT £5, "DETERMINANT="; det
890 REMark check multiply
900 FOR i=1 TO n
910 PRINT £5, i;"        ";
920 FOR j=1 TO n
930 p=0
940 FOR k=1 TO n
950 p=p+l(i,k)*l(j,k)
960 NEXT k
970 PRINT £5, p;"   ";
980 NEXT j
990 PRINT £5
1000 NEXT i
1010 REMark DATA FOR HONEYDEW TEST
1020 DATA 0, 4452.496,5990.488,6812.086, 3423.528
1030 DATA 0, 3466.685, 7837.82, 6349.145
1040 DATA 0, 5506.923, 5893.098
1050 DATA 0, 3827.922
1060 DATA 0
```

```
MATRIX TO BE FACTORISED
1.982472E7      2.184638E7      2.398907E6      -4.383189E6

2.184638E7      3.588595E7      2.598213E7      6.438943E6

2.398907E6      2.598213E7      4.640452E7      2.173604E7

-4.383189E6     6.438943E6      2.173604E7      1.172054E7


1       4452.496   0  0  0
2       4906.547   3436.822  0  0
3       538.778    6790.748  5.564957  0
4       -984.4341  3278.935  .6120891  4.106968
DETERMINANT=3.497387E8
1    1.982472E7    2.184638E7    2.398907E6    -4.383189E6
2    2.184638E7    3.588595E7    2.598213E7    6.438943E6
3    2.398907E6    2.598213E7    4.640458E7    2.173604E7
4    -4.383189E6   6.438943E6    2.173604E7    1.172054E7
```

CONVFT

Much of the code for this program is the same as that described in FFT. The main differences between the two programs are as follows:

The real and imaginary parts of the first function, $f(x)$, whose values are entered at line 120, are stored in the variables U(P) and V(P) (lines 350 and 355).

After printing the real and imaginary parts of the transform (line 327), the program transfers to line 120 to receive the second set of values $g(z)$. The transform values are again printed at line 327 and the program transfers to line 370 where the convolution product for the real and imaginary parts are formed. The inverse transform of this product is then calculated by transforming to line 140 and the real and imaginary parts — that is, the convolution results — are printed at line 435.

```
95 REM CONVFT
96 REM CONVOLUTION OF TWO FUNCTIONS
97 SW = 0
100 PRINT "ENTER L WHERE N=2^L"
105 INPUT L
110 N = 2 ^ L
115 DIM X(N),Y(N),F(N),G(N),U(N),V(N)
120 PRINT "ENTER DATA AS X Y PAIRS"
125 FOR I = 1 TO N
130 INPUT X(I),Y(I)
135 NEXT I
140 M = N / 2
145 R = 2 ^ (L - 1)
150 PI = 3.141593
155 H = 1
160 A = 2 * PI / N
165 FOR P = 1 TO L
170 B = INT ((H - 1) / R)
175 GOSUB 1000
180 B = A * B
190 I = SIN (B)
195 J = COS (B)
200 FOR Q = 1 TO M
205 K = H + M
210 C = X(K) * J + Y(K) * I
215 D = Y(K) * J - X(K) * I
220 X(K) = X(H) - C
225 Y(K) = Y(H) - D
230 X(H) = X(H) + C
235 Y(H) = Y(H) + D
240 H = H + 1
245 NEXT Q
250 H = H + M
```

```
255  IF H < = N GOTO 170
260  H = 1
265  R = R / 2
270  M = M / 2
275  NEXT P
276  IF SW = 2 GOTO 400
277  PRINT "TRANSFORM VALUES"
280  FOR P = 1 TO N
285  B = P - 1
290  GOSUB 1000
295  H = B + 1
320  F(P) = X(H) / N
325  G(P) = Y(H) / N
327  PRINT F(P);" ";G(P)
330  NEXT P
335  IF SW = 1 GOTO 370
340  IF SW = 2 GOTO 400
345  FOR P = 1 TO N
350  U(P) = F(P)
355  V(P) = G(P)
360  NEXT P
365  SW = 1: GOTO 120
370  FOR P = 1 TO N
375  X(P) = U(P) * F(P) - V(P) * G(P)
380  Y(P) = - V(P) * F(P) - U(P) * G(P)
385  NEXT P
390  SW = 2
395  GOTO 140
400  PRINT "CONVOLUTION RESULT"
405  FOR P = 1 TO N
410  B = P - 1
415  GOSUB 1000
420  H = B + 1
425  F(P) = X(H)
430  G(P) = Y(H)
435  PRINT F(P);" ";G(P)
440  NEXT P
445  END
1000  K = 0
1005  FOR I = 1 TO L
1010  J = INT (B / 2)
1015  K = B + INT (2 * K - 2 * J)
1020  B = J
1025  NEXT I
1030  B = K
1035  RETURN


]RUN
ENTER L WHERE N=2^L
?3
ENTER DATA AS X Y PAIRS
?1,0
?1,0
?1,0
?1,0
?0,0
?0,0
```

```
?0,0
?0,0
TRANSFORM VALUES
.5 0
.124999952 -.301776669
0 0
.124999955 -.0517767127
0 0
.125000002 .0517766824
0 0
.125000088 .301776699
ENTER DATA AS X Y PAIRS
?0.5,0
?0.5,0
?0.5,0
?0.5,0
?0,0
?0,0
?0,0
?0,0
TRANSFORM VALUES
.25 0
.0624999759 -.150888334
0 0
.0624999777 -.0258883563
0 0
.0625000012 .0258883412
0 0
.062500044 .150888349
CONVOLUTION RESULT
.0625000033 -4.35611583E-08
.125000003 -1.82437415E-08
.187500004 -1.05239332E-08
.24999999 1.70985004E-10
.187499993 4.35611583E-08
.124999993 1.82437415E-08
.0624999928 1.05239332E-08
1.03318598E-08 -1.70985004E-10
```

DETER

The calculation of the determinant of a square matrix is an essential procedure. To multiply out the terms for a matrix of dimensions greater than four or so is prohibitively long so that the method used here is that of pivotal condensation. The largest term is found and using it as the 'pivot', the other terms in the column containing it are reduced to zero. The determinant is thus reduced by one in dimension and the process is repeated. Any standard text on matrix algebra will explain the rules involved.

```
100 REM read in data
110 REM order of matrix
120 N=3
130 FOR I=1 TO N
140 FOR J=1 TO N
150 READ D(I,J)
160 NEXT J
170 NEXT I
180 DATA 7,11,4
190 DATA 13,15,10
```

```
200 DATA 3,9,6
220 REM Program name DETER
230 REM Calculation of determinant
240 REM by pivotal condensation
250 REM Det. of order N is in D(N,N). Result in S.
260 REM Matrix D is destroyed. Uses A(N*N).
270 REM and I,J,IV,IP,IQ; and S,A,B
280 REM begins
290 S=1
300 B=0
310 REM find largest element=pivot - D(IP,IQ)
320 FOR I=1 TO N
330 FOR J=1 TO N
340 A=ABS(D(I,J))
350 IF A<B THEN 390
360 IP=I
370 IQ=J
380 B=A
390 NEXT J
400 NEXT I
410 REM multiply by largest element
420 S=S*D(IP,IQ)*(-1)^(IP+IQ)
430 IF S=0 THEN 680
440 IV=1
450 REM reduce other terms by subtr. cols. of pivot
460 FOR I=1 TO N
470 IF I=IP THEN 560
480 FOR J=1 TO N
490 IF J=IQ THEN 550
500 D(I,J)=D(I,J)-D(IP,J)/D(IP,IQ)*D(I,IQ)
510 A(IV)=D(I,J)
520 IV=IV+1
530 REM run out non-zero elements
540 REM into determinant of lower order
550 NEXT J
560 NEXT I
570 N=N-1
580 IF N=0 THEN 680
590 IV=1
600 FOR I=1 TO N
610 FOR J=1 TO N
620 D(I,J)=A(IV)
630 IV=IV+1
640 NEXT J
650 NEXT I
660 REM repeat until det. has order 1
670 GOTO 300
680 PRINT "DETERMINANT=";S
690 STOP
```

```
DETERMINANT=-240
```

DISC

A dot matrix printer can be used for quite elaborate drawing. It is a matter of studying the printer manual to discover the necessary commands and the computer manual to find out how to transmit them. Both of these stages are often tedious. We given an example here for the Epson FX-80 printer which involves using the print image commands and also those for winding the printer backwards and forwards. With such a program as a basis, arbitrary molecules can be drawn as overlapping discs.

```
100 REM program name DISC
110 REM to draw discs
120 REM centre is at a,b. Radius r
130 REM initialise printer
140 LPRINT CHR$(27);"@";
150 REM set line height
160 LPRINT CHR$(27);"3";CHR$(24);
170 REM line length infinite (MBASIC)
180 WIDTH LPRINT 255
190 DEFINT I-N
200 REM number of discs = NP
210 NP=5
220 REM read in coords of centres and radii
230 FOR I=1 TO NP
240 READ A(I),B(I),R(I)
250 IA(I)=INT(A(I))
260 IB(I)=INT(B(I))
270 IR(I)=INT(R(I))
280 NEXT I
290 DATA 50,50,30
300 DATA 50,150,30
310 DATA 150,50,30
320 DATA 150,150,30
330 DATA 100,100,60
340 REM measures in dots
350 DIM M(480)
360 REM number of lines at 8 dots per line
370 ILL=25
380 REM number of dots horizontally
390 N=200
400 IW(8)=1
410 IW(7)=2
420 IW(6)=4
430 IW(5)=8
440 IW(4)=16
450 IW(3)=32
460 IW(2)=64
470 IW(1)=128
480 FOR I=1 TO NP
490 IS(I)=IR(I)*IR(I)-IA(I)*IA(I)
500 NEXT I
510 PRINT "K"
520 FOR K=1 TO NP
530 FOR IL=1 TO ILL
540 IDX=0
550 FOR JX=1 TO 8
560 IX=JX+8*IL-8
570 ID=IS(K)-IX*IX+2*IA(K)*IX
580 IF ID<0 THEN GOTO 680
590 IDX=1
600 IT=INT(SQR(ID))
610 IYB=INT(.828*(IB(K)-IT))
620 IYS=INT(.828*(IB(K)+IT))
630 IF IYB<0 THEN IYB=0
640 IF IYS>N THEN IYS=N
650 FOR I=IYB TO IYS
660 M(I)=M(I)+IW(JX)
670 NEXT I
680 NEXT JX
690 IF IDX=0 THEN 780
700 REM print dot pattern
```

```
710 LPRINT CHR$(27);"*";CHR$(0);
720 LPRINT CHR$(N MOD 256);CHR$(INT(N/256));
730 FOR I=1 TO N
740 IN=M(I)
750 M(I)=0
760 LPRINT CHR$(IN);
770 NEXT I
780 LPRINT
790 NEXT IL
800 REM rewind
810 FOR I=1 TO ILL
820 LPRINT; CHR$(27);"j";CHR$(24);
830 NEXT I
840 NEXT K
850 REM re-initialise printer
860 LPRINT CHR$(27);"@";
870 STOP
```

FFT

This program calculates the Fourier transform of a set of values using the successive-doubling method.

Line 100 prompts for the value of L where L is defined as $N = 2^L$. N corresponds to the total number of data points.

The data is entered as X, Y pairs at line 130 and so if the data consists of real values these will correspond to the elements of the X array with the Y array values being entered as zero. Since the resulting transform of a real array is complex — that is, consists of both real and imaginary parts — the X array will eventually contain the real values and the Y array the imaginary values of the transform. The constants used in the program are established in lines 140-160 inclusive.

The implementation of eqns (10.11)-(10.14) for each successive argument is carried out in lines 165-275 with the bit reversal procedure taking place in the sub-routine beginning at line 1000. (Note that bit manipulation is effected by dividing or multiplying by powers of 2 since each bit corresponds to a particular value of 2^m where m is an integer.) The exponential multiplication is carried out on the terms corresponding to F_{odd} — that is, $X(K)$, $Y(K)$ in accordance with eqns (10.13) and (10.14). Since the resulting transform consists of values (real and imaginary) corresponding to the arguments in the re-ordered array, we must re-order our values back to the original array to get the correct transform. This is carried out in lines 280–340 using the bit reversal routine and storing the result in the arrays F and G. F and G correspond to the final real and imaginary parts of the transform and these are squared and added to form the amplitude spectrum at line 350.

```
95  REM  FFT
100  PRINT "ENTER L WHERE N=2^L"
105  INPUT L
110  N = 2 ^ L
115  DIM X(N),Y(N),F(N),G(N)
120  PRINT "ENTER DATA AS X Y PAIRS"
125  FOR I = 1 TO N
130  INPUT X(I),Y(I)
135  NEXT I
140  M = N / 2
145  R = 2 ^ (L - 1)
150  PI = 3.141593
155  H = 1
160  A = 2 * PI / N
165  FOR P = 1 TO L
170  B =  INT ((H - 1) / R)
175  GOSUB 1000
180  B = A * B
190  I =  SIN (B)
195  J =  COS (B)
200  FOR Q = 1 TO M
205  K = H + M
210  C = X(K) * J + Y(K) * I
215  D = Y(K) * J - X(K) * I
220  X(K) = X(H) - C
225  Y(K) = Y(H) - D
230  X(H) = X(H) + C
235  Y(H) = Y(H) + D
240  H = H + 1
245  NEXT Q
250  H = H + M
255  IF H <  = N GOTO 170
260  H = 1
265  R = R / 2
270  M = M / 2
275  NEXT P
277  PRINT "TRANSFORM VALUES"
280  FOR P = 1 TO N
285  B = P - 1
290  GOSUB 1000
295  H = B + 1
320  F(P) = X(H) / N
325  G(P) = Y(H) / N
330  PRINT F(P);" ";G(P)
340  NEXT P
343  PRINT "SPECTRAL VALUES"
345  FOR I = 1 TO N
350  X(I) =  SQR (F(I) ^ 2 + G(I) ^ 2)
355  PRINT "X(I)=";X(I)
```

```
360  NEXT I
365  END
1000 K = 0
1005 FOR I = 1 TO L
1010 J = INT (B / 2)
1015 K = B + INT (2 * K - 2 * J)
1020 B = J
1025 NEXT I
1030 B = K
1035 RETURN
```

```
]RUN
ENTER L WHERE N=2^L
?4
ENTER DATA AS X Y PAIRS
?0,0
?0,0
?0,0
?0,0
?0,0
?0,0
?1,0
?1,0
?1,0
?1,0
?0,0
?0,0
?0,0
?0,0
?0,0
?0,0
TRANSFORM VALUES
.25 0
-.222179128 -.0441941383
.150888364 .0624999785
-.0661413006 -.0441941708
1.45519127E-11 -1.45519144E-11
.0295295957 .0441941461
-.0258883484 -.0624999683
8.79077581E-03 .0441941317
2.91038305E-11 0
8.79076164E-03 -.0441941814
-.0258883423 .0625000146
.029529605 -.0441942104
1.45519178E-11 1.45519144E-11
-.0661412294 .0441941735
.150888327 -.0625000248
-.22217908 .0441942494
SPECTRAL VALUES
X(I)=.25
X(I)=.226531867
X(I)=.163320378
X(I)=.0795474482
X(I)=2.05795152E-11
X(I)=.0531518539
X(I)=.0676494835
X(I)=.0450599492
X(I)=2.91038305E-11
X(I)=.0450599951
X(I)=.067649524
X(I)=.0531519125
X(I)=2.05795173E-11
X(I)=.0795473897
X(I)=.163320361
X(I)=.226531842
```

FIBONACCI

An example of a very simple graphics program to draw almost close-packed circles in a spiral using polar coordinates. The centres of the circles lie on a spiral at distances from the centre increasing as the square root of the natural numbers and at angular intervals of the Golden Angle which is $360°$ divided by τ (the Golden Number = 1.618. . .).

```
100 REMark program name "FIBONACCI"
110 CLS
120 POINT 50,50
130 t=(1+SQRT(5))/2
140 th=2*PI/t
150 REMark adjustable scale factor
160 k=3
170 rds=k*t/2
180 FOR i=1 TO 400
190 r=k*SQRT(i)
200 theta=th*i
210 x=r*SIN(theta)
220 y=r*COS(theta)
230 CIRCLE x+50,y+50,rds
240 NEXT i
```

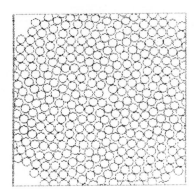

FWT

This program is based on the successive-doubling FFT algorithm of Cooley *et al.* (1969) with the trigonometric terms set equal to one.

The program prompts for the value LN where LN is defined by $N = 2^{LN}$. N is the total number of data points. The values for the function $f(x)$ are entered at line 125. The procedure for implementing eqns (11.33) and (11.34) is similar to that used in the FFT algorithm previously described with the notable exception that the trigonometric terms are set equal to unity. The results are printed in the loop (lines 315–333). Note that the input data consists of single values rather than pairs of values as specified in the algorithm FFT.

```
95   REM   FWT
100  PRINT "ENTER LN WHERE N=2^LN
105  INPUT LN
110  N = 2 ^ LN
115  DIM F(N)
120  FOR I = 1 TO N
125  PRINT "ENTER F(X)": INPUT F(I)
130  NEXT I
135  PI = 3.14159
140  NV = N / 2
145  NM = N - 1
150  J = 1
155  FOR I = 1 TO NM
160  IF I > = J GOTO 190
165  T = F(J)
170  F(J) = F(I)
175  F(I) = T
190  K = NV
195  IF K > = J GOTO 220
200  J = J - K
210  K = K / 2
215  GOTO 195
220  J = J + K
223  NEXT I
225  FOR L = 1 TO LN
230  LE = 2 ^ L
235  L1 = LE / 2
260  FOR J = 1 TO L1
265  FOR I = J TO N STEP LE
270  IP = I + L1
275  T = F(IP)
280  F(IP) = F(I) - T
285  F(I) = F(I) + T
302  NEXT I
312  NEXT J: NEXT L
315  FOR I = 1 TO N
320  F(I) = F(I) / N
330  PRINT F(I)
333  NEXT I
335  END
```

```
]RUN
ENTER LN WHERE N=2^LN
?3
ENTER F(X)
?1
ENTER F(X)
?2
ENTER F(X)
?3
ENTER F(X)
?4
ENTER F(X)
?0
ENTER F(X)
?0
ENTER F(X)
?0
ENTER F(X)
```

```
?0
1.25
1.25
-.5
-.5
-.25
-.25
0
0
```

GENDAT

This program calculates data distributions which are used subsequently to illustrate sampling and spectral amplitude characteristics of the Fourier transform. The distributions are the Gaussian function $y = A \exp - B^2(x - c)^2$; the Poisson distribution $y = A^x \exp(-A)/x!$; and the sine curve $y = A \sin 2\pi x$. Line 110 prompts for the total number of points over which we require to evaluate the function.

Line 125 prompts for the distribution required: G for a Gaussian, P for a Poisson or S for a sine wave. The sine wave is calculated in the loop specified by lines 155-175 with a value for the amplitude being specified at line 150. The Gaussian is calculated in the loop (lines 190-210) with the values A, B, C being requested at line 185. The Poisson distribution is calculated in lines 225-1020 where the sub-program (lines 1000-1020) calculates the values for $x!$. The value for 'A' is specified at line 220.

```
95 REM  GENDAT
100 REM  PROGRAM GENERATES DATA SETS
105 TWOPI = 6.28318
110 PRINT "GIVE NO. OF PTS IN SET"
115 INPUT N:M = N - 1
120 PRINT "SPECIFY DATA TYPE"
125 PRINT "GAUSSIAN(G),POISSON(P),SINEWAVE(S)"
130 INPUT DT$
135 IF DT$ = "G" GOTO 180
140 IF DT$ = "P" GOTO 215
145 PRINT "SPECIFY AMPLITUDE"
150 INPUT A
155 FOR I = 0 TO M
160 X = I / M
165 Y = A * SIN (TWOPI * X)
170 PRINT "X=";X;" ";"Y=";Y
175 NEXT I: GOTO 265
180 PRINT " SPECIFY A,B,C VALUES"
185 INPUT A,B,C
190 FOR I = 0 TO M
195 X = I / M
200 B1 = (B ^ 2) * (X - C) ^ 2:Y = A * EXP ( - B1)
205 PRINT "X=";X;" ";"Y=";Y
210 NEXT I: GOTO 265
```

```
215  PRINT "SPECIFY A VALUE"
220  INPUT A
225  Y = EXP ( - A):X = 0
230  PRINT "X=";X;"  ";"Y=";Y
235  FOR I = 2 TO N
240  J = I - 1:X = J
245  GOSUB 1000
250  Y = (A ^ X) * EXP ( - A) / B
255  PRINT "X=";X;"  ";"Y=";Y
260  NEXT I
265  END
1000 B = 1
1005 FOR J = 2 TO X
1010 B = B * J
1015 NEXT J
1020 RETURN
```

```
]RUN
GIVE NO. OF PTS IN SET
?16
SPECIFY DATA TYPE
GAUSSIAN(G),POISSON(P),SINEWAVE(S)
?G
 SPECIFY A,B,C VALUES
?2,2.5,0
X=0   Y=2
X=.0666666667   Y=1.94520895
X=.133333333   Y=1.78967863
X=.2   Y=1.55760157
X=.266666667   Y=1.28236078
X=.333333333   Y=.998703576
X=.4   Y=.735758882
X=.466666667   Y=.512751513
X=.533333333   Y=.33802663
X=.6   Y=.210798449
X=.666666667   Y=.124353047
X=.733333333   Y=.0693933712
X=.8   Y=.0366312777
X=.866666667   Y=.018291894
X=.933333333   Y=8.64047891E-03
X=1   Y=3.86090827E-03
```

```
]RUN
GIVE NO. OF PTS IN SET
?16
SPECIFY DATA TYPE
GAUSSIAN(G),POISSON(P),SINEWAVE(S)
?P
SPECIFY A VALUE
?10
X=0   Y=4.53999298E-05
X=1   Y=2.26999649E-04
X=2   Y=2.26999649E-03
X=3   Y=7.56665497E-03
X=4   Y=.0189166374
X=5   Y=.0378332748
X=6   Y=.0630554581
X=7   Y=.0900792258
```

```
X=8   Y=.112599032
X=9   Y=.125110036
X=10  Y=.125110036
X=11  Y=.113736396
X=12  Y=.0947803303
X=13  Y=.0729079463
X=14  Y=.0520771045
X=15  Y=.0347180697
```

GENINV

The calculation of the generalised inverse of a matrix (singular or non-singular, square or rectangular) is an essential system function rarely provided. We give an iterative method for finding this inverse. An expression is iterated until the trace (the sum of the diagonal terms) of a matrix becomes an integer, to a degree of accuracy which we may specify. The trace then gives the rank of the matrix (which must be an integer). As a starting model for the inverse of a matrix we take its transpose multiplied by a sufficiently small constant. If iteration does not converge, then this constant may be made smaller.

```
100 REM program name GENINV
110 REM generalised inverse of X(n,m)
120 READ N,M
130 DIM H(N),P(M),Q(N)
140 DIM X(N,M),Y(M,N),W(M,N),Z(N,N),A(M,N)
150 REM matrix entered in x and returned in y
160 REM read in matrix
170 FOR I=1 TO N
180 FOR J=1 TO M
190 READ X(I,J)
200 LET W(J,I)=X(I,J)
210 REM w is transpose of x
220 NEXT J
230 REM r.h.s. of equation
240 READ H(I)
250 NEXT I
260 LET K=0
270 FOR I=1 TO N
280 FOR J= 1 TO N
290 FOR L=1 TO N
300 LET Z(I,J)=Z(I,J)+X(I,L)*W(L,J)
310 NEXT L
320 LET K=K+ABS(Z(I,J))
330 NEXT J
340 NEXT I
350 LET K=1/K
360 LPRINT "K=";K
370 REM small constant
380 LET D=.00001
390 LPRINT "constant for integral trace=";D
400 LPRINT "trace+2*n"
410 FOR I=1 TO M
420 FOR J=1 TO N
430 REM first approximation to inverse
440 LET Y(I,J)=K*W(I,J)
450 NEXT J
460 NEXT I
470 FOR I=1 TO N
```

```
480 FOR J=1 TO N
490 LET Z(I,J)=0
500 FOR L=1 TO M
510 LET Z(I,J)=Z(I,J)+X(I,L)*Y(L,J)
520 NEXT L
530 NEXT J
540 NEXT I
550 REM trace = t
560 LET T=0
570 FOR I= 1 TO N
580 LET Z(I,I)=Z(I,I)-2
590 LET T=T+Z(I,I)
600 NEXT I
610 LPRINT 2*N+T
620 FOR I=1 TO M
630 FOR J= 1 TO N
640 LET W(I,J)=0
650 FOR L=1 TO N
660 LET W(I,J)=W(I,J)+Y(I,L)*Z(L,J)
670 NEXT L
680 NEXT J
690 NEXT I
700 FOR I= 1 TO M
710 FOR J= 1 TO N
720 LET Y(I,J)=-W(I,J)
730 NEXT J
740 NEXT I
750 IF ABS(T-INT(T)-1)<D THEN 780
760 IF ABS(T-INT(T))<D THEN 780
770 GOTO 470
780 REM repeat until t is an integer
790 FOR I= 1 TO M
800 FOR J=1 TO M
810 LET A(I,J)=0
820 FOR L=1 TO N
830 LET A(I,J)=A(I,J)+Y(I,L)*X(L,J)
840 NEXT L
850 NEXT J
860 NEXT I
870 LPRINT "rank of matrix="; 2*N+T
880 REM remove next statement for full printout
890 GOTO 1170
900 LPRINT "generalised inverse"
910 LPRINT
920 FOR I= 1 TO M
930 FOR J=1 TO N
940 LPRINT Y(I,J),
950 NEXT J
960 LPRINT
970 LPRINT
980 NEXT I
990 REM checking procedure
1000 LPRINT
1010 LPRINT "original matrix"
1020 FOR I= 1 TO N
1030 FOR J= 1 TO M
1040 LPRINT X(I,J),
1050 NEXT J
1060 LPRINT
1070 LPRINT
1080 NEXT I
1090 LPRINT
1100 LPRINT "products"
1110 FOR I=1 TO N
1120 FOR J=1 TO N
1130 LPRINT Z(I,J),
1140 NEXT J
```

```
1150 LPRINT
1160 NEXT I
1170 LPRINT "solutions to equations"
1180 FOR I=1 TO M
1190 LET P(I)=0
1200 FOR J=1 TO N
1210 LET P(I)=P(I)+Y(I,J)*H(J)
1220 NEXT J
1230 LPRINT I, P(I)
1240 NEXT I
1250 LPRINT "number of equations=";N
1260 LPRINT "number of unknowns=";M
1270 LPRINT "calculated and observed r.h.s."
1280 FOR I=1 TO N
1290 LET Q(I)=0
1300 FOR J= 1 TO M
1310 LET Q(I)=Q(I)+X(I,J)*P(J)
1320 NEXT J
1330 LPRINT Q(I),H(I),H(I)-Q(I)
1340 NEXT I
1350 REM test data
1360 DATA 4,4
1370 DATA 5,4,3,2,1496
1380 DATA 4,2,6,3,1175
1390 DATA 3,1,7,5,958
1400 DATA 2,3,5,1,861
1410 DATA 2,3,5,1,861
1420 DATA 2,3,5,1,862
1430 DATA 1,2,3,4,500
```

```
K= .0011655
constant for integral trace= .00001
trace+2*n
 .282051
 .498261
 .795824
1.10459
1.36429
1.66213
2.06907
2.51502
2.89665
3.1903
3.42252
3.66956
3.89082
3.98808
3.99986
4
```

```
rank of matrix= 4
generalised inverse
```

7.12713E-08	.666667	-.333333	-.333333
.290323	-.72043	.247312	.344086
-.225806	.301075	-.11828	.139785
.258064	-.677419	.516129	-.0645161

```
original matrix
 5              4              3              2

 4              2              6              3

 3              1              7              5

 2              3              5              1


products
-1              -3.57628E-07   0              -1.04308E-07
 5.96046E-08    -1             0              -1.04308E-07
 0              0              -1             -8.9407E-08
 1.78814E-07    5.96046E-08    0              -1
solutions to equations
 1              177
 2              121
 3              23
 4              28.9999
number of equations= 4
number of unknowns= 4
calculated and observed r.h.s.
 1496           1496           -3.66211E-04
 1175           1175           -2.44141E-04
 958            958            1.83105E-04
 861            861            1.2207E-04
```

GREEK

The Epson FX-80 printer has effectively become an industry standard for the
code required for printing with an 8 dot matrix printer. It can readily be re-
programmed so as to produce the arbitrary symbols used in mathematics. We
give here a version of the Greek alphabet plus various mathematical symbols.
The new characters are downloaded into the italic font of the printer and are
called as if italics were to be used. It is necessary to direct that the 2K buffer
should be freed for this purpose. Once the program has been run the characters
remain available until the printer is re-initialised. The printer manual gives fuller
explanations.

```
100 REM change type font
110 REM program name GREEK
120 REM copyright A.L.Mackay,
130 REM Birkbeck College, London
140 REM Microsoft Basic for Nascom II micro
150 REM and Epson FX-80 matrix printer
160 REM set infinite line width
170 WIDTH LPRINT 255
180 REM put Greek characters into Italic set
190 REM initialise printer
200 REM
210 REM type style condensed enlarged
220 REM
230 REM set left margin
240 REM
250 REM copy original characters
```

```
260 LPRINT CHR$(27);":";CHR$(0);CHR$(0);CHR$(0);
270 REM select download set
280 LPRINT CHR$(27);"%";CHR$(1);CHR$(0);
290 DEFINT I-N
300 DIM L(11)
310 REM number of characters to be re-defined
320 N=44
330 DIM A$(44)
340 DATA a,L,X,F,g,G,l,x,f,D
350 DATA d,m,P,h,e,z,n,p,W,j
360 DATA Q,q,r,w,i,k,y,t,u,U
370 DATA c,b,J,R,S,H,s,£,$
380 DATA %,&,[,!,]
390 FOR I= 1 TO N
400 READ A$(I)
410 NEXT I
420 REM if there are descenders in new chars.
430 REM then attribute is 11, otherwise 139
440 REM list of attribute chars.
450 DIM M(44)
460 FOR I=1 TO N
470 READ M(I)
480 NEXT I
490 DATA 139,139,139,139,11,139,139,139,11,139
500 DATA 139,11,139,139,139,11,139,139,139,11
510 DATA 139,139,11,11,139,139,139,139,139,139
520 DATA 11,11,139,139,139,139,139,139,139
530 DATA 139,139,139,139,139
540 REM redefine characters
550 FOR I=1 TO N
560 LPRINT CHR$(27);"&";CHR$(0);
570 LPRINT CHR$(128+ASC(A$(I)));CHR$(128+ASC(A$(I)));
580 LPRINT CHR$(M(I));
590 FOR J=1 TO 11
600 READ L(J)
610 NEXT J
620 FOR J=1 TO 11:LPRINT CHR$(L(J));:NEXT J
630 NEXT I
640 REM data for N characters
650 REM 11 items for each
660 REM data for GREEK
670 DATA 0,28,0,34,0,34,20,8,20,34,0
680 DATA 0,2,4,8,16,32,64,128,112,14,0
690 DATA 0,2,0,146,0,146,0,146,0,128,0
700 DATA 0,24,36,2,76,16,100,128,72,48,0
710 DATA 0,32,64,135,0,138,84,40,64,128,0
720 DATA 0,6,24,96,128,0,128,0,128,0,128
730 DATA 0,130,0,132,64,40,16,8,4,2,0
740 DATA 0,40,85,128,85,0,85,34,64,32,0
750 DATA 0,16,41,2,86,16,100,128,40,16,0
760 DATA 0,2,4,10,16,34,64,130,112,14,0
770 REM
780 DATA 0,0,76,162,16,130,16,130,76,0,0
790 DATA 0,3,12,48,68,0,4,8,52,64,0
800 DATA 0,6,24,96,128,0,128,6,152,96,128
810 DATA 0,34,0,34,20,8,20,34,64,2,0
820 DATA 0,20,42,0,42,0,34,20,0,0
830 DATA 0,1,0,177,8,66,136,66,140,64,0
840 DATA 0,32,18,12,2,0,4,8,16,48,0
850 DATA 0,34,4,56,0,32,0,32,28,34,0
860 DATA 0,128,120,5,128,127,128,5,120,128,0
870 DATA 0,64,12,48,0,64,3,76,48,0,0
880 REM
890 DATA 0,28,34,80,130,16,130,20,136,112,0
900 DATA 0,12,18,8,34,8,34,8,36,24,0
910 DATA 0,7,24,32,4,64,4,64,8,48,0
920 DATA 0,96,16,0,11,20,104,128,16,96,0
```

```
930 DATA 0,0,0,0,0,60,2,0,2,0,0
940 DATA 0,0,62,0,8,16,36,0,2,0,0
950 DATA 0,32,12,48,2,0,2,4,56,0,0
960 DATA 0,32,0,60,2,32,2,32,0,32,0
970 DATA 0,28,34,0,2,28,2,0,34,28,0
980 DATA 0,50,72,2,132,0,132,2,72,50,0

990 REM
1000 DATA 0,48,72,1,68,1,68,1,70,32,0
1010 DATA 0,127,128,4,160,4,160,4,88,0,0
1020 DATA 0,6,56,192,16,0,16,6,56,192,0
1030 DATA 0,6,24,96,144,0,144,0,144,96,0
1040 DATA 0,2,0,134,64,170,0,146,0,128,0
1050 DATA 0,2,132,72,32,24,36,2,64,128,0
1060 DATA 0,12,16,34,0,34,0,52,8,32,0
1070 DATA 0,76,146,0,162,0,162,0,68,56,0
1080 DATA 129,66,129,36,137,16,129,0,129,0,195
1090 REM
1100 DATA 64,48,64,8,70,1,64,6,72,48,64
1110 DATA 4,2,1,6,8,48,64,0,64,0,64
1120 DATA 0,48,76,131,0,128,0,128,64,48,0
1130 DATA 0,0,0,128,96,24,6,1,0,0,0
1140 DATA 0,6,2,1,0,1,0,193,50,12,0
1145 REM remove next statement to test
1150 GOTO 1440
1160 LPRINT "Shift to Greek by esc4 gk esc5"
1170 LPRINT "English to Greek:"
1180 LPRINT
1190 LPRINT "ABCDEFGHIJKLMNOPQRSTUVWXYZ"
1200 LPRINT CHR$(27);"4";
1210 LPRINT "ABCDEFGHIJKLMNOPQRSTUVWXYZ"
1220 LPRINT CHR$(27);"5";
1230 LPRINT "abcdefghijklmnopqrstuvwxyz"
1240 LPRINT CHR$(27);"4";
1250 LPRINT "abcdefghijklmnopqrstuvwxyz"
1260 LPRINT CHR$(27);"5";
1270 LPRINT
1280 LPRINT "Greek to English:"
1290 LPRINT
1300 LPRINT CHR$(27);"4";
1310 LPRINT "ABGDEZJQIKLMNXOPRSTYFHWU"
1320 LPRINT CHR$(27);"5";
1330 LPRINT;"ABGDEZJQIKLMNXOPRSTYFHWU"
1340 LPRINT CHR$(27);"4";
1350 LPRINT "abgdezjqiklmnxoprstyfhwuc"
1360 LPRINT CHR$(27);"5";
1370 LPRINT "abgdezjqiklmnxoprstyfhwuc"
1380 LPRINT
1390 LPRINT "Special characters"
1400 LPRINT "£";" ";"$";" ";"%";" ";"&";" ";"[";" ";"!";" ";"]
"
1410 LPRINT CHR$(27);"4";
1420 LPRINT "£";" ";"$";" ";"%";" ";"&";" ";"[";" ";"!";" ";"]
"
1430 LPRINT CHR$(27);"5";
1440 STOP
```

```
Shift to Greek by esc4 gk esc5
English to Greek:

ABCDEFGHIJKLMNOPQRSTUVWXYZ
ΑΒΓ ΔΕ Φ ΓΧΙ ΗΚ ΛΜΝΟ ΠΘΡΣΤ ΩΥ ΨΞΥΖ
abcdefghijklmnopqrstuvwxyz
αβςδεφγχιηκλμνοπθρστων γξυς
```

Greek to English:

ABΓΔEZHΘIKΛMNΞOΠPΣTYΦXΨΩ
ABGDEZJQIKLMNXOPRSTYFHWU
αβγδεζηθικλμνξοπρστυφχψως
abgdezjqiklmnxoprstyfhwuc

Special characters
£ $ % & [!]
ə Σ ∇ √ ∩ \ ∪

HADAMARD

This program determines the Hadamard transform for $N = 8$.

The values of $b_i(x)$ and $b_i(u)$ in the transformation kernel are stored in the array B(I, J). The values for $g(x, u)$ are calculated in the loop (lines 155–165). Each value for $f(x)$ is requested at line 190 and the final values for the transform are calculated in the loop (lines 210–220) with the results being printed at line 230.

```
95  REM  HADAMARD
100  DIM F(8),G(8,8),B(8,8),H(8)
105  FOR I = 0 TO 7
110  FOR J = 0 TO 7
115  B(I,J) = 0
120  NEXT J: NEXT I
125  B(0,1) = 1:B(1,2) = 1:B(0,3) = 1:B(1,3) = 1
130  B(2,4) = 1:B(0,5) = 1:B(2,5) = 1:B(1,6) = 1
135  B(2,6) = 1:B(0,7) = 1:B(1,7) = 1:B(2,7) = 1
140  FOR X = 0 TO 7
145  FOR U = 0 TO 7
150  G(X,U) = 0
155  FOR I = 0 TO 2
160  G(X,U) = (B(I,X) * B(I,U)) + G(X,U)
165  NEXT I
166  G(X,U) = ( - 1) ^ G(X,U)
175  NEXT U: NEXT X
180  FOR X = 0 TO 7
185  PRINT "ENTER FX"
190  INPUT F(X)
195  NEXT X
200  FOR U = 0 TO 7
205  H(U) = 0
210  FOR X = 0 TO 7
215  H(U) = H(U) + F(X) * G(X,U)
220  NEXT X
225  H(U) = H(U) / 8
230  PRINT U;" ";H(U)
235  NEXT U
240  END
```

```
]RUN
ENTER FX
?1
ENTER FX
?2
ENTER FX
?3
ENTER FX
?4
ENTER FX
?0
ENTER FX
?0
ENTER FX
?0
ENTER FX
?0
0  1.25
1  -.25
2  -.5
3  0
4  1.25
5  -.25
6  -.5
7  0
```

IMAGES

This program enables the user to generate a series of images constructed by
filling the various elements of a 16 × 16 array. The program then calculates the
Fourier transform of the shape and displays the corresponding amplitudes
scaled between 0 and 9. The program is illustrated in relation to an image
formed by the letter T. Lines 105–140 show how the letter is formed by filling
the elements of the T matrix using the value 5 at certain points within the two-
dimensional matrix. The matrix is then displayed on a line-by-line basis (lines
145–155). Since we are using the one-dimensional FFT, the elements of the two-
dimensional T matrix are stored in the one-dimensional matrix DZ on a line-
by-line basis ready for input into the FFT.

   The transform is calculated in lines 1140–4035. A 'count' JC is established
to show the number of times or lines the transform has calculated and is printed
at line 2077. The real and imaginary parts of each successive transform are
stored in the arrays RZ and IZ (lines 235–255). Since we are dealing with a two-
dimensional array, remember that we also have to take the transform by columns
as well as rows (lines 285–315). The final values for the amplitudes are stored in
array DZ at line 320. Finally, for display purposes the results are re-stored in the
array T (lines 328–332). Since the largest value is the origin term stored in T
(1.9), a scaling value SC is evaluated at line 350 which, when applied to each
value of the T matrix, scales all the amplitudes to lie between 0 and 9. The final
matrix of transformed amplitudes is then displayed in lines 145–155.

```
0  REM  IMAGES
5  REM  FT OF T AND SQUARE
00  DIM T(17,17),X(17),Y(17),F(17),G(17),RZ(300),IZ(300),DZ(300)
03  N = 16:L = 4
05  FOR I = 7 TO 8
10  FOR J = 6 TO 11
15  T(I,J) = 5
20  NEXT J: NEXT I
25  FOR I = 9 TO 11
30  FOR J = 8 TO 9
35  T(I,J) = 5
40  NEXT J: NEXT I
45  FOR I = 1 TO 16
50  PRINT T(I,1);" ";T(I,2);" ";T(I,3);" ";T(I,4);" ";T(I,5);" ";T(I,6);" ";T(I,7);" ";T(I,8);" ";T(I,9);" ";T(I,10);" ";T(I,
11);" ";T(I,12);" ";T(I,13);" ";T(I,14);" ";T(I,15);" ";T(I,16)
55  NEXT I
56  IF JC = 17 GOTO 390
65  FOR I = 1 TO 16
70  FOR J = 1 TO 16
75  M = J + 16 * (I - 1)
80  DZ(M) = T(I,J)
85  NEXT J: NEXT I
90  MM = 1:NN = 16:IB = 0:JC = 0
95  FOR M = MM TO NN
200  IB = IB + 1
05  X(IB) = DZ(M)
210  Y(IB) = 0
215  NEXT M
220  JC = JC + 1
225  GOSUB 1140
230  J = JC
235  FOR P = 1 TO 16
240  RZ(J) = F(P)
245  IZ(J) = G(P)
250  J = J + 16
255  NEXT P
260  MM = MM + 16:NN = NN + 16:IB = 0
265  IF MM > 256 GOTO 280
275  GOTO 195
280  MM = 1:NN = 16:IB = 0:JC = 0
285  FOR M = MM TO NN
290  IB = IB + 1
295  X(IB) = RZ(M)
300  Y(IB) = IZ(M)
305  NEXT M
310  JC = JC + 1
315  GOSUB 1140
316  P = 1
318  FOR JJ = 1 TO 16
320  DZ(JJ) = SQR (F(P) ^ 2 + G(P) ^ 2)
326  P = P + 1
327  NEXT JJ
328  FOR JJ = 9 TO 16
329  JA = JJ - 8
330  T(JC,JA) = DZ(JJ)
331  T(JC,JJ) = DZ(JA)
332  NEXT JJ
333  MM = MM + 16:NN = NN + 16:IB = 0
```

```
335  IF MM > 256 GOTO 350
340  GOTO 285
350  SC = 9 / T(1,9)
355  FOR JC = 1 TO 16
360  FOR JJ = 1 TO 16
365  T(JC,JJ) = SC * T(JC,JJ)
370  T(JC,JJ) =  INT (T(JC,JJ))
375  NEXT JJ: NEXT JC
380  GOTO 145
390  END
1140 M = N / 2
1145 R = 2 ^ (L - 1)
1150 PI = 3.141593
1155 H = 1
1160 A = 2 * PI / N
1165  FOR P = 1 TO L
1170 B =  INT ((H - 1) / R)
1175  GOSUB 4000
1180 B = A * B
1190 I =  SIN (B)
1195 J =  COS (B)
2000 FOR Q = 1 TO M
2005 K = H + M
2010 C = X(K) * J + Y(K) * I
2015 D = Y(K) * J - X(K) * I
2020 X(K) = X(H) - C
2025 Y(K) = Y(H) - D
2030 X(H) = X(H) + C
2035 Y(H) = Y(H) + D
2040 H = H + 1
2045  NEXT Q
2050 H = H + M
2055  IF H < = N GOTO 1170
2060 H = 1
2065 R = R / 2
2070 M = M / 2
2075  NEXT P
2077  PRINT "TRANSFORM";" ";JC
2080  FOR P = 1 TO N
2085 B = P - 1
2090  GOSUB 4000
2095 H = B + 1
3020 F(P) = X(H) / N
3025 G(P) = Y(H) / N
3030  REM
3040  NEXT P
4000 K = 0
4005  FOR I = 1 TO L
4010 J =  INT (B / 2)
4015 K = B +  INT (2 * K - 2 * J)
4020 B = J
4025  NEXT I
4030 B = K
4035  RETURN
```

```
]RUN
0 0 0 0 0 0 0 0 0 0 0 0 0 0 0 0
0 0 0 0 0 0 0 0 0 0 0 0 0 0 0 0
0 0 0 0 0 0 0 0 0 0 0 0 0 0 0 0
0 0 0 0 0 0 0 0 0 0 0 0 0 0 0 0
0 0 0 0 0 0 0 0 0 0 0 0 0 0 0 0
0 0 0 0 0 0 0 0 0 0 0 0 0 0 0 0
0 0 0 0 0 5 5 5 5 5 5 0 0 0 0 0
0 0 0 0 0 5 5 5 5 5 5 0 0 0 0 0
0 0 0 0 0 5 5 5 5 5 5 0 0 0 0 0
0 0 0 0 0 0 0 0 0 0 0 0 0 0 0 0
0 0 0 0 0 0 0 0 0 0 0 0 0 0 0 0
0 0 0 0 0 0 0 0 0 0 0 0 0 0 0 0
0 0 0 0 0 0 0 0 0 0 0 0 0 0 0 0
0 0 0 0 0 0 0 0 0 0 0 0 0 0 0 0
0 0 0 0 0 0 0 0 0 0 0 0 0 0 0 0
0 0 0 0 0 0 0 0 0 0 0 0 0 0 0 0
TRANSFORM 1
TRANSFORM 2
TRANSFORM 3
TRANSFORM 4
TRANSFORM 5
TRANSFORM 6
TRANSFORM 7
TRANSFORM 8
TRANSFORM 9
TRANSFORM 10
TRANSFORM 11
TRANSFORM 12
TRANSFORM 13
TRANSFORM 14
TRANSFORM 15
TRANSFORM 16
TRANSFORM 1
TRANSFORM 2
TRANSFORM 3
TRANSFORM 4
TRANSFORM 5
TRANSFORM 6
TRANSFORM 7
TRANSFORM 8
TRANSFORM 9
TRANSFORM 10
TRANSFORM 11
TRANSFORM 12
TRANSFORM 13
TRANSFORM 14
TRANSFORM 15
TRANSFORM 16
3 2 1 0 2 5 7 8 9 8 7 5 2 0 1 2
2 2 0 0 2 4 5 6 7 6 5 4 2 0 0 2
0 0 0 0 0 1 2 2 2 2 1 0 0 0 0 0
0 0 0 0 0 0 0 1 0 0 0 0 0 0 0 0
0 0 0 0 0 1 1 2 2 2 1 0 0 0 0 0
0 0 0 0 0 0 0 0 0 0 0 0 0 0 0 0
0 0 0 0 0 0 1 1 1 0 0 0 0 0 0 0
0 0 0 0 0 0 1 1 1 1 0 0 0 0 0 0
0 0 0 0 0 0 0 0 0 0 0 0 0 0 0 0
0 0 0 0 0 0 1 1 1 1 0 0 0 0 0 0
0 0 0 0 0 0 1 1 1 0 0 0 0 0 0 0
0 0 0 0 0 0 0 0 0 0 0 0 0 0 0 0
0 0 0 0 0 1 1 2 2 2 1 1 0 0 0 0
0 0 0 0 0 0 0 1 0 0 0 0 0 0 0 0
0 0 0 0 0 1 2 2 2 2 1 0 0 0 0 0
2 2 0 0 2 4 5 6 7 6 5 4 2 0 0 2
```

```
]RUN
0 0 0 0 0 0 0 0 0 0 0 0 0 0 0 0
0 0 0 0 0 0 0 0 0 0 0 0 0 0 0 0
0 0 0 0 0 0 0 0 0 0 0 0 0 0 0 0
0 0 0 0 0 0 0 0 0 0 0 0 0 0 0 0
0 0 0 0 0 0 0 0 0 0 0 0 0 0 0 0
0 0 0 0 0 0 0 0 0 0 0 0 0 0 0 0
0 0 0 0 0 5 5 5 5 5 5 0 0 0 0 0
0 0 0 0 0 5 5 5 5 5 5 0 0 0 0 0
0 0 0 0 0 0 0 5 5 0 0 0 0 0 0 0
0 0 0 0 0 0 0 5 5 0 0 0 0 0 0 0
0 0 0 0 0 0 0 5 5 0 0 0 0 0 0 0
0 0 0 0 0 0 0 0 0 0 0 0 0 0 0 0
0 0 0 0 0 0 0 0 0 0 0 0 0 0 0 0
0 0 0 0 0 0 0 0 0 0 0 0 0 0 0 0
0 0 0 0 0 0 0 0 0 0 0 0 0 0 0 0
0 0 0 0 0 0 0 0 0 0 0 0 0 0 0 0
TRANSFORM 1
TRANSFORM 2
TRANSFORM 3
TRANSFORM 4
TRANSFORM 5
TRANSFORM 6
TRANSFORM 7
TRANSFORM 8
TRANSFORM 9
TRANSFORM 10
TRANSFORM 11
TRANSFORM 12
TRANSFORM 13
TRANSFORM 14
TRANSFORM 15
TRANSFORM 16
TRANSFORM 1
TRANSFORM 2
TRANSFORM 3
TRANSFORM 4
TRANSFORM 5
TRANSFORM 6
TRANSFORM 7
TRANSFORM 8
TRANSFORM 9
TRANSFORM 10
TRANSFORM 11
TRANSFORM 12
TRANSFORM 13
TRANSFORM 14
TRANSFORM 15
TRANSFORM 16
1 0 1 3 3 3 5 7 9 7 5 3 3 3 1 0
0 0 1 2 2 2 4 6 7 6 4 2 2 2 1 0
0 0 0 1 0 0 2 3 4 3 2 0 0 1 0 0
0 0 0 0 1 2 2 2 1 2 2 2 1 0 0 0
0 0 0 0 1 2 2 1 0 1 2 2 1 0 0 0
0 0 0 0 0 1 1 1 1 1 1 1 0 0 0 0
0 0 0 0 0 0 1 1 1 1 0 0 0 0 0 0
0 0 0 0 0 0 0 1 1 1 0 0 0 0 0 0
0 0 0 0 0 0 0 0 0 0 0 0 0 0 0 0
0 0 0 0 0 0 0 1 1 1 0 0 0 0 0 0
0 0 0 0 0 0 0 1 1 1 0 0 0 0 0 0
0 0 0 0 0 1 1 1 1 1 1 1 0 0 0 0
0 0 0 0 1 2 2 1 0 1 2 2 1 0 0 0
0 0 0 0 1 2 2 2 1 2 2 2 1 0 0 0
0 0 0 1 0 0 2 3 4 3 2 0 0 1 0 0
0 0 1 2 2 2 4 6 7 6 4 2 2 2 1 0
```

INVERT

A standard textbook program, as given for example by Poole and Borchers (1979), to find the inverse of a non-singular matrix by the Gauss-Jordan elimination method. It entails division by a diagonal element and if this is zero the program will fail. It is then necessary to exchange rows and columns and to try again.

```
1000 REM program name INVERT
1010 REM program for matrix inversion
1020 DEFINT I-N
1030 REM dimensions of matrix
1040 READ N
1050 DIM A(N,N), B(N,N)
1060 DIM A1(N,N)
1070 REM matrix input in a
1080 PRINT
1090 PRINT "Matrix input"
1100 FOR I= 1 TO N
1110 FOR J= 1 TO N
1120 READ A(I,J)
1130 A1(I,J)=A(I,J)
1140 PRINT A(I,J);
1150 NEXT J
1160 PRINT
1170 NEXT I
1180 REM matrix inverse in b
1190 FOR I= 1 TO N
1200 B(I,I)=1
1210 NEXT I
1220 FOR J= 1 TO N
1230 FOR I=J TO N
1240 IF A(I,J) <> 0 THEN 1280
1250 NEXT I
1260 PRINT "zero"
1270 GOTO 1500
1280 FOR K= 1 TO N
1290 S=A(J,K)
1300 A(J,K)=A(I,K)
1310 A(I,K)=S
1320 S=B(J,K)
1330 B(J,K)=B(I,K)
1340 B(I,K)=S
1350 NEXT K
1360 T= 1/A(J,J)
1370 FOR K= 1 TO N
1380 A(J,K)=T*A(J,K)
1390 B(J,K)=T*B(J,K)
1400 NEXT K
1410 FOR L= 1 TO N
1420 IF L=J THEN 1480
1430 T=-A(L,J)
1440 FOR K= 1 TO N
1450 A(L,K)=A(L,K)+T*A(J,K)
1460 B(L,K)=B(L,K)+T*B(J,K)
1470 NEXT K
1480 NEXT L
1490 NEXT J
1500 PRINT
1510 PRINT "INVERSE"
1520 FOR I=1 TO N
1530 FOR J= 1 TO N
1540 PRINT B(I,J);"   ";
1550 NEXT J
```

```
1560 PRINT
1570 NEXT I
1580 REM order of matrix
1590 DATA 3
1600 REM elements of matrix by rows
1610 DATA 1,-2,3
1620 DATA 2,-5,10
1630 DATA -1,2,-2
1650 PRINT
1660 PRINT "Test by multiplication"
1670 FOR I=1 TO N
1680 PRINT
1690 FOR J=1 TO N
1700 PR=0
1710 FOR K=1 TO N
1720 PR=PR + A1(I,K)*B(K,J)
1730 NEXT K
1740 PRINT PR;
1750 NEXT J
1760 NEXT I
```

```
Matrix input
  1  -2   3
  2  -5  10
 -1   2  -2

INVERSE
 10    -2     5
  6    -1     4
  1     0     1

Test by multiplication

  1   0   0
  0   1   0
  0   0   1
```

JACOBI

The Jacobi method diagonalises a real symmetric matrix by rotating the axes.
That is, it reduces two off-diagonal terms to zero, looking successively for the
greatest and rotating appropriately. The diagonal terms are then the eigenvalues
of the matrix and the final rotation matrix contains the eigenvectors. The method
is sometimes slow but is the most reliable method and does not fail if there are
equal or zero eigenvalues.

```
90 REM program name JACOBI
100 REM Jacobi diagonalisation of sym. matrix
110 REM order=N, Matrix in A(N,N)
120 READ N
130 DIM A(N,N),T(N,N),R(N,N),W(N,N),U(N,N)
140 PI=3.141592653500003£
150 LPRINT "Original matrix"
160 FOR I=1 TO N
170 R(I,I)=1
180 LPRINT
190 FOR J=1 TO N
200 READ A(I,J)
```

```
210 LPRINT A(I,J);
220 NEXT J
230 LPRINT
240 NEXT I
250 LPRINT
260 LPRINT "Largest term. Rotation in deg."
270 REM find largest off-diagonal term
280 LET A9=0
290 FOR I=1 TO N
300 FOR J=1 TO N
310 IF I=J THEN GOTO 360
320 IF A9>ABS(A(I,J)) THEN GOTO 360
330 A9=ABS(A(I,J))
340 I9=I
350 J9=J
360 NEXT J
370 NEXT I
380 IF ABS(A9) < .00001 THEN GOTO 960
390 FOR I=1 TO N
400 FOR J= 1 TO N
410 T(I,J)=0
420 NEXT J
430 T(I,I)=1
440 NEXT I
450 IF ABS(A(I9,I9)-A(J9,J9))>.0000001 THEN GOTO 480
460 T9=PI*.25
470 GOTO 500
480 T9=.5*ATN(2*A(I9,J9)/(A(I9,I9)-A(J9,J9)))
490 LPRINT A9,T9*180/PI
500 T(I9,I9)=COS(T9)
510 T(J9,J9)=COS(T9)
520 T(J9,I9)=SIN(T9)
530 T(I9,J9)=-SIN(T9)
540 FOR I=1 TO N
550 FOR J=1 TO N
560 U(I,J)=T(J,I)
570 NEXT J
580 NEXT I
590 REM MAT(W)=A*T
600 FOR I=1 TO N
610 FOR J=1 TO N
620 PR=0
630 FOR K=1 TO N
640 PR=PR+A(I,K)*T(K,J)
650 NEXT K
660 W(I,J)=PR
670 NEXT J
680 NEXT I
690 REM MAT(A)=U*W
700 FOR I=1 TO N
710 FOR J=1 TO N
720 PR=0
730 FOR K=1 TO N
740 PR=PR+U(I,K)*W(K,J)
750 NEXT K
760 A(I,J)=PR
770 NEXT J
780 NEXT I
790 REM MAT(W)=R*T
800 FOR I=1 TO N
810 FOR J=1 TO N
820 PR=0
830 FOR K=1 TO N
840 PR=PR+R(I,K)*T(K,J)
850 NEXT K
860 W(I,J)=PR
870 NEXT J
880 NEXT I
```

```
890 FOR I=1 TO N
900 FOR J=1 TO N
910 R(I,J)=W(I,J)
920 NEXT J
930 NEXT I
940 GOTO 270
950 LPRINT
960 LPRINT "Diagonalised matrix"
970 LPRINT "Eigenvalues"
980 FOR I=1 TO N
990 LPRINT I,A(I,I)
1000 NEXT I
1010 LPRINT
1020 LPRINT "Matrix to be applied for diagonalisation"
1030 LPRINT "Rows are Eigenvectors"
1040 FOR I=1 TO N
1050 FOR J=1 TO N
1060 LPRINT R(J,I),
1070 NEXT J
1080 LPRINT
1090 NEXT I
1100 DATA 7
1110 DATA 0,1,3,4,3,1,1
1120 DATA 1,0,1,3,4,3,1
1130 DATA 3,1,0,1,3,4,1
1140 DATA 4,3,1,0,1,3,1
1150 DATA 3,4,3,1,0,1,1
1160 DATA 1,3,4,3,1,0,1
1170 DATA 1,1,1,1,1,1,0
1180 STOP
```

Original matrix

0	1	3	4	3	1	1
1	0	1	3	4	3	1
3	1	0	1	3	4	1
4	3	1	0	1	3	1
3	4	3	1	0	1	1
1	3	4	3	1	0	1
1	1	1	1	1	1	0

Largest term.	Rotation in deg.
4	22.5
5.22625	23.6329
2.44426	-11.1295
2.13908	-43.0819
2.71061	32.0583
.713445	3.28638
.521831	-1.61876
.0751449	.344978
.030957	.321378
.0299223	-.285745
5.41455E-03	.645374
8.52902E-04	3.91545E-03
5.7149E-04	5.45732E-03
5.68208E-04	.0677203
5.61107E-04	-2.48035E-03

```
 2.69729E-05   2.80007E-04
Diagonalised matrix
Eigenvalues
 1              -6.00001
 2               2.54615E-07
 3              -6
 4              -8.43255E-08
 5               0
 6               12.4808
 7              -.480741

Matrix to be applied for diagonalisation
Rows are Eigenvectors
 .577351
 .288676
-.288675
-.577351
-.288675
 .288676
 8.38695E-08

-.62321
 .494217
-.23623
 .107237
-.23623
 .494217
 3.44589E-08

 0
 .5
 .5
 0
-.5
-.5
 0

 .33408
-.0758278
-.440676
 .698928
-.440676
-.0758278
 6.29234E-08

 0
 .5
-.5
 0
 .5
-.5
 0

 .400606
 .400606
 .400606
 .400606
 .400606
 .400606
 .192588

-.0786235
-.0786236
-.0786235
-.0786236
-.0786235
-.0786236
 .98128
```

JULESZ

This is another exercise in the use of the Epson FX-80 matrix plotter.

```
100 REM program name "JULESZ"
110 REM to generate a stereoscopic pair of random elements
120 REM REF: Bela Julesz, "Foundations of Cyclopean
130 REM        Perception", Univ. of Chicago Press, 1971
140 REM        and Scientific American, Feb. 1965. pp.38-48
150 REM version for EPSON FX-80 printer
160 LPRINT CHR$(27);"@";
170 LPRINT CHR$(27);"3";CHR$(24);
180 REM number of squares per side
190 N=68
200 REM number of squares in displaced element
210 M=14
220 REM required by EPSON printer
230 WIDTH LPRINT 255
240 DEFINT J
250 DIM J(N,N)
260 FOR I=1 TO N
270 X=RND
280 FOR K=1 TO N
290 J(I,K)=INT(2*RND)
300 NEXT K: NEXT I
310 C=0
320 C=C+1
330 IF C=3 THEN GOTO 450
340 FOR K=1 TO N
350 LPRINT CHR$(27);"L";CHR$((M*N)MOD 256);CHR$(INT(M*N/256));
360 FOR I=1 TO N
370 IF J(K,I)=0 THEN GOSUB 470
380 IF J(K,I)=1 THEN GOSUB 520
390 NEXT I
400 LPRINT
410 NEXT K
420 LPRINT: LPRINT: LPRINT
430 GOSUB 570
440 GOTO 320
450 LPRINT CHR$(27);"@";
460 STOP
470 REM subroutine fill
480 FOR L=1 TO M
490 LPRINT CHR$(255);
500 NEXT L
510 RETURN
520 REM subroutine empty
530 FOR L=1 TO M
540 LPRINT CHR$(0);
550 NEXT L
560 RETURN
570 REM subroutine to shift part of square
580 FOR K=3*N/4 TO N/4 STEP -1
590 FOR I=N/4 TO 3*N/4
600 J(K+1,I)=J(K,I)
610 NEXT I
620 NEXT K
630 RETURN
```

LAPLACE

This is the simplest example of a 'molecular dynamics' program which traces the state of a mechanical system from moment to moment. It was originally written for a Sinclair Spectrum microcomputer but can be adapted for any other system supporting graphics.

```
100 REMark program name LAPLACE
110 REMark traces satellite orbits under mutual
111 REMark gravitational forces
120 REMark number of bodies=N
130 n=3
140 REMark set gravitaional constant G
150 G=1
160 REMark set time interval for plotting positions
```

```
170 dt=.2
180 DIM X(5): DIM Y(5): REMark positions x,y
190 DIM p(5): DIM q(5): REMark previous posns. x,y
200 DIM r(5): DIM s(5): REMark forces in x,y directions
210 DIM m(5): REMark masses
220 REMark read in mass of each body
230 FOR i=1 TO n
240 READ m(i)
250 NEXT i
260 DATA 10000, 100, 100
270 REMark read in initial x,y posns. of each body
280 FOR i=1 TO n
290 READ X(i),Y(i)
300 NEXT i
310 DATA 128, 83
320 DATA 178, 83
330 DATA 188, 83
340 REMark read in velocities in x,y, directions
350 FOR i=1 TO n
360 READ vx,vy
370 p(i)=X(i)-vx*dt: q(i)=Y(i)-vy*dt
380 NEXT i
390 DATA 0, -.267
400 DATA 0, 15.7
410 DATA 0,11
420 REMark iterated part of program begins
430 REMark sum forces on each body in x and y dirns.
440 FOR i=1 TO n
450 r(i)=0: s(i)=0
460 FOR j=1 TO n
470 IF i=j THEN GO TO 520
480 dx=X(i)-X(j): dy=Y(i)-Y(j)
490 r=sqr(dx*dx+dy*dy)
500 t=-G*m(i)*m(j)/r/r
510 r(i)=r(i)+t*dx/r: s(i)=s(i)+t*dy/r
520 NEXT j
530 NEXT i
540 REMark compute future positions
550 FOR i=1 TO n
560 xf=2*X(i)-p(i)+dt*dt*r(i)/m(i)
570 yf=2*Y(i)-q(i)+dt*dt*s(i)/m(i)
580 IF X(i)>255 THEN GO TO 630
590 IF X(i)<0 THEN GO TO 630
600 IF Y(i)>175 THEN GO TO 630
610 IF Y(i)<0 THEN GO TO 630
620 plot X(i), Y(i)
630 REMark update positions
640 p(i)=X(i): q(i)=Y(i)
650 X(i)=xf: Y(i)=yf
660 NEXT i
670 GO TO 420
```

ONEDFT

This program calculates the one-dimensional Fourier transform of a set of sampled values of FX together with the spectral components.

Line 100 prompts for the total number of sampled values. These are entered individually in the loop defined between lines 125 and 135 inclusive and stored in variable FX(I). The real and imaginary parts of the transform are calculated as

$$FX(J) \cos(-2\pi(I-1)(J-1)) \text{ and } FX(J) \sin(-2\pi(I-1)(J-1))$$

respectively and stored in variables A(J), B(J). These values are normalised at line 185 and printed at line 190.

```
90  REM  ONEDFT
95  REM  PROGRAM COMPUTES FOURIER TRANSFORM IN ONE DIMENSION
100  PRINT " ENTER NO. OF POINTS "
105  INPUT N
110  DIM FX(N),A(N),B(N)
115  TWOPI = 6.28318
120  PRINT " ENTER VALUES FOR FX "
125  FOR I = 1 TO N
130  INPUT FX(I)
135  NEXT I
137  PRINT "REAL AND IMAG PARTS"
140  FOR I = 1 TO N
145  A(0) = 0:B(0) = 0
150  FOR J = 1 TO N
155  K = I - 1:L = J - 1
160  A(J) = FX(J) * COS ( - TWOPI * K * L / N) + A(L)
165  B(J) = FX(J) * SIN ( - TWOPI * K * L / N) + B(L)
170  NEXT J
185  A(N) = A(N) / N:B(N) = B(N) / N
190  PRINT A(N);"    ";B(N)
195  NEXT I
200  END
```

```
]RUN
 ENTER NO. OF POINTS
?4
 ENTER VALUES FOR FX
?1
?2
?3
?4
REAL AND IMAG PARTS
2.5        0
-.500003314        .499998011
-.5        -5.30307804E-06
-.499990057        -.500005967
```

PACK

This is an example of an iterative program to pack N circles on the surface of a sphere so that the minimum distance between their centres may be a maximum. It will show the difficulties of such a task, such as the way in which the discs will get stuck in a local minimum and the configuration will not refine to the globally best packing. Many much more sophisticated programs encounter this basic physical difficulty and many stratagems, such as applying a small random

movement when the configuration gets stuck, may be adopted to try to deal
with it. The program uses the formula for the great circle distance on a sphere.

```
100 REMark program name PACK
110 OPEN £5, ser1
120 CLS
130 cycle=1
140 REMark pack points on a sphere
150 REMark number of points
160 N=20
170 REMark initial target distance
180 drec=46.36/180*PI
190 REMark target distance in degrees
200 dlim=46.75
210 dlim=dlim*PI/180
220 REMark theta and phi coordinates
230 DIM T(N), F(N)
240 REMark corections
250 DIM DT(N), DF(N)
260 GO TO 370
270 REMark start with random points
280 OPEN_NEW £6, mdv1_DATA
290 FOR i=1 TO N
300 F(i)=2*PI*RND
310 T(i)=ASIN(2*RND-1)
320 IF T(i)<0 THEN T(i)=T(i)+PI
330 PRINT £6, T(i)
340 PRINT £6, F(i)
350 NEXT i
360 CLOSE £6
370 DEFine FuNction DIST(n1,n2)
380 LET ANS=COS(T(n1))*COS(T(n2))
390 LET ANS=ANS+SIN(T(n1))*SIN(T(n2))*COS(F(n1)-F(n2))
400 LET ANS=ACOS(ANS)
410 RETurn ANS
420 END DEFine
430 DEFine FuNction CORRT(n1,n2)
440 LET ANS=COS(T(n1))*SIN(T(n2))*COS(F(n1)-F(n2))
450 LET ANS=ANS-SIN(T(n1))*COS(T(n2))
460 LET ANS=SIN(d)/ANS
470 RETurn ANS
480 END DEFine
490 DEFine FuNction CORRF(n1,n2)
500 LET ANS=SIN(T(n1))*SIN(T(n2))*SIN(F(n1)-F(n2))
510 LET ANS=-SIN(d)/ANS
520 RETurn ANS
530 END DEFine
540 OPEN £6, mdv1_DATA
550 FOR i=1 TO N
560 INPUT £6, T(i): INPUT £6, F(i)
570 NEXT i
580 CLOSE £6
590 REMark begin adjustment
600 REMark adjust 25 times with
610 REMark decreasing correction
620 FOR fac=.12 TO 1E-2 STEP -5E-3
630 FOR i=1 TO N
640 DT(i)=0
650 DF(i)=0
660 NEXT i
670 dmin=PI
680 FOR i=2 TO N
690 FOR j=1 TO i-1
700 d=DIST(i,j)
```

```
710 IF d<dmin THEN dmin=d
720 d=d-dlim
730 REMark correct only those closer to limit
740 IF d>0 THEN GO TO 800
750 DT(i)=DT(i)+d*CORRT(i,j)
760 DT(j)=DT(j)+d*CORRT(j,i)
770 dff=d*CORRF(i,j)
780 DF(i)=DF(i)+dff
790 DF(j)=DF(j)-dff
800 NEXT j
810 NEXT i
820 REMark record best configuration
830 IF dmin+1E-2>drec THEN GO TO 920
840 PRINT
850 PRINT £5, "dmin=";dmin*180/PI
860 PRINT £5, "I, Theta, Phi"
870 FOR l=1 TO N
880 PRINT £5,l,T(l),F(l)
890 NEXT l
900 drec=dmin
910 PRINT
920 REMark apply corrections
930 FOR i=1 TO N
940 T(i)=T(i)-fac*ATAN(10*DT(i))
950 F(i)=F(i)-fac*ATAN(10*DF(i))
960 NEXT i
970 REMark plot stereogram
980 CLS £2
990 CLS £1
1000 PRINT "cycle=";cycle;"   ";
1010 cycle=cycle+1
1020 PRINT "limit=";dlim/PI*180
1030 PRINT "dmin=";dmin*180/PI;"   ";
1040 PRINT £5, dmin*180/PI;"   ";
1050 CIRCLE 50,50,40
1060 CIRCLE £2,50,50,40
1070 FOR i=1 TO N
1080 IF ABS(T(i))>PI/2 THEN GO TO 1150
1090 r=40*TAN(T(i)/2)
1100 xp= 50+r*COS(F(i)): yp= 50+r*SIN(F(i))
1110 POINT xp,yp
1120 CURSOR xp,yp,0,0
1130 PRINT i;
1140 GO TO 1200
1150 r=40*TAN((PI-ABS(T(i)))/2)*T(i)/ABS(T(i))
1160 xp=50+ COS(F(i))*r: yp=50+r*SIN(F(i))
1170 POINT £2, xp,yp
1180 CURSOR #2
1190 PRINT £2, i;
1200 NEXT i
1210 NEXT fac
1220 REMark apply random variation
1230 w=.2
1240 w=w/180*PI
1245 GO TO 1300
1250 FOR i=1 TO N
1260 T(i)=T(i)+w*(.5-RND)
1270 F(i)=F(i)+w*(.5-RND)
1280 NEXT i
1290 PRINT "random=";w/PI*180
1300 REMark repeat
1310 REMark write new data to file
1320 DELETE mdv1_DATA
1330 OPEN_NEW £6, mdv1_DATA
1340 FOR i=1 TO N
1350 PRINT £6, T(i)
1360 PRINT £6, F(i)
```

```
1370 NEXT i
1380 CLOSE £6
1390 GO TO 540
```

```
46.54014   46.54269   46.54507   46.54729   46.54937   46.5513   46
.55311   46.55479   46.55634   46.55779   46.55912   46.56034   46.
56147   46.56249   46.56342   46.56426   46.56501   46.56566   46.5
6623   46.56672   46.56712   46.56745
```

PENROSE

The Penrose pattern or tiling has become of great interest. It consists of two
kinds of tiles, each a rhombus, one of vertex angle 72° and the other of vertex
angle 144°. All sides are equal. Each tile can be subdivided into copies of the
two tiles which are smaller by a factor of $\tau = 1.618$. The program does this and
plots successive generations on the screen. The version is for the BBC computer
with a colour monitor (or TV). The number of steps of subdivision which can be
plotted is severely limited by the resolution of the screen.

```
 10   REM PROGRAM NAME "PEN"
 20  MODE 1
 30  COLOUR 1
 40  VDU 19,1,2,0,0,0
 50  COLOUR 2
 60  VDU 19,2,6,0,0,0
 70  TAU=.5*(1+SQR(5))
 80  T1=TAU-1
 90  T2=2-TAU
100 REM SHIFT ORIGIN
110   VDU 29,1100;512;
120REM SCALE CONSTANT
130  C=550
140C72=COS(2*PI/5)
150C36=COS(PI/5)
160C18=COS(PI/10)
170C54=SIN(PI/5)
180   DIM POSX(140),POSY(140)
190DIM H(5)
200 DIM X(4),Y(4)
210 FOR I=1 TO 4
220FOR J=1 TO 5
230READ H(J)
240NEXT J
250POSX(I)=FNX(C)
260POSY(I)=FNY(C)
270NEXT I
280 PROCEXFOL1(0,4)
290DATA 0,0,0,0,0
300DATA 0,0,1,0,0
310DATA 0,0,1,1,0
320DATA 0,0,0,1,0
330 REM ZERO GENERATION
340 REM
```

```
 350 PROCPLOT(1,1,1)
 360 REM
 370 PROCDELAY(2)
 380 REM FIRST GENERATION
 390 FOR I=5 TO 24 STEP 4
 400 IF I<17 THEN D=1
 410 IF I>16 THEN D=2
 420 PROCPLOT(I,1,D)
 430 NEXT I
 440 PROCDELAY(2)
 450 REM SECOND GENERATION
 460   PROCEXFOL1(4,24)
 470   PROCEXFOL1(8,44)
 480   PROCEXFOL1(12,64)
 490   PROCEXFOL2(16,84)
 500   PROCEXFOL2(20,100)
 510 REM PLOT
 520 REM LENGTH OF LONGER DIAG.
 530 DS=C*C72*2
 540 FOR I=25 TO 116 STEP 4
 550 D=1
 560 IF ABS(FND(I,I+2)-DS) <1E-6 THEN D=2
 570 PROCPLOT(I,1,D)
 580 NEXT I
 590 END
 600 REM
 610 DEF PROCPLOT(K,S,D)
 620 REM K=FIRST POINT OF RHOMB
 630 REM S=SCALE
 640 REM D=COLOUR
 650 LOCAL I,X1,Y1
 660 X1=S*POSX(K)
 670 Y1=S*POSY(K)
 680 MOVE X1,Y1
 690 MOVE X1,Y1
 700 GCOLO,D
 710 FOR I=1 TO 3
 720 PLOT 85, S*POSX(K+I),S*POSY(K+I)
 730 NEXT I
 740 PLOT 85, X1, Y1
 750 GCOL 0,0
 760 FOR I=1 TO 3
 770 PLOT 5,S*POSX(K+I),S*POSY(K+I)
 780 NEXT I
 790 PLOT 5, X1,Y1
 800 ENDPROC
 810 DEF PROCEXFOL1(K,P)
 820 REM BEGIN AFTER POINT K
 830 REM PUT RESULTS AFTER POINT P
 840 LOCAL I,X,Y
 850 FOR I=1 TO 4
 860 X(I)=POSX(K+I)
 870 Y(I)=POSY(K+I)
 880 NEXT I
 890 REM TYPE1
 900 POSX(P+1)=X(3)
 910   POSY(P+1)=Y(3)
 920   POSX(P+2)=X(3)*T2+X(4)*T1
 930   POSY(P+2)=Y(3)*T2+Y(4)*T1
 940   POSX(P+3)=X(1)*T1+X(3)*T2
 950    POSY(P+3)=Y(1)*T1+Y(3)*T2
 960   POSX(P+4)=X(2)*T1+X(3)*T2
 970    POSY(P+4)=Y(2)*T1+Y(3)*T2
 980 REM TYPE 1
 990 POSX(P+5)=X(2)
1000   POSY(P+5)=Y(2)
1010   POSX(P+6)=X(1)*T1+X(3)*T2
1020   POSY(P+6)=Y(1)*T1+Y(3)*T2
```

```
1030 POSX(P+7)=X(1)
1040  POSY(P+7)=Y(1)
1050   POSX(P+8)=X(2)+X(1)*(1-T1)-X(3)*T2
1060    POSY(P+8)=Y(2)+Y(1)*(1-T1)-Y(3)*T2
1070 REM TYPE 1
1080 POSX(P+9)=X(4)
1090  POSY(P+9)=Y(4)
1100 POSX(P+10)=X(1)*(1-T1)+X(4)-X(3)*T2
1110  POSY(P+10)=Y(1)*(1-T1)+Y(4)-Y(3)*T2
1120POSX(P+11)=X(1)
1130 POSY(P+11)=Y(1)
1140POSX(P+12)=X(1)*T1+X(3)*T2
1150 POSY(P+12)=Y(1)*T1+Y(3)*T2
1160REM TYPE 2
1170POSX(P+13)=X(2)
1180  POSY(P+13)=Y(2)
1190POSX(P+14)=X(2)*(1+T1)-X(1)*T1
1200  POSY(P+14)=Y(2)*(1+T1)-Y(1)*T1
1210POSX(P+15)=X(2)*T1+X(3)*T2
1220 POSY(P+15)=Y(2)*T1+Y(3)*T2
1230POSX(P+16)=X(1)*T1+X(3)*T2
1240 POSY(P+16)=Y(1)*T1+Y(3)*T2
1250REM TYPE 2
1260POSX(P+17)=X(4)
1270 POSY(P+17)=Y(4)
1280POSX(P+18)=X(1)*T1+X(3)*T2
1290 POSY(P+18)=Y(1)*T1+Y(3)*T2
1300POSX(P+19)=X(4)*T1+X(3)*T2
1310 POSY(P+19)=Y(4)*T1+Y(3)*T2

1320POSX(P+20)=X(4)*(1+T1)-X(1)*T1
1330 POSY(P+20)=Y(4)*(1+T1)-Y(1)*T1
1340 ENDPROC
1350REM EXFOLIATION OF TYPE 2
1360 DEF PROCEXFOL2(K,P)
1370 LOCAL I,X,Y
1380 FOR I=1 TO 4
1390    X(I)=POSX(K+I)
1400    Y(I)=POSY(K+I)
1410 NEXT I
1420REM TYPE 1
1430POSX(P+1)=X(2)
1440 POSY(P+1)=Y(2)
1450POSX(P+2)=X(3)*T1+X(2)*T2
1460 POSY(P+2)=Y(3)*T1+Y(2)*T2
1470POSX(P+3)=X(1)
1480 POSY(P+3)=Y(1)
1490POSX(P+4)=X(1)*(1+T1)-X(4)*T1
1500 POSY(P+4)=Y(1)*(1+T1)-Y(4)*T1
1510REM TYPE 1
1520POSX(P+5)=X(4)
1530 POSY(P+5)=Y(4)
1540POSX(P+6)=X(1)*(1+T1)-X(2)*T1
1550 POSY(P+6)=Y(1)*(1+T1)-Y(2)*T1
1560POSX(P+7)=X(1)
1570 POSY(P+7)=Y(1)
1580POSX(P+8)=X(3)*T1+X(4)*T2
1590 POSY(P+8)=Y(3)*T1+Y(4)*T2
1600REM TYPE 2
1610POSX(P+9)=X(3)
1620 POSY(P+9)=Y(3)
1630POSX(P+10)=X(1)
1640 POSY(P+10)=Y(1)
1650POSX(P+11)=X(3)*T1+X(2)*T2
1660 POSY(P+11)=Y(3)*T1+Y(2)*T2
1670POSX(P+12)=X(3)*(1+T1)+X(2)*T2-X(1)
1680 POSY(P+12)=Y(3)*(1+T1)+Y(2)*T2-Y(1)
1690REM TYPE 2
```

```
1700POSX(P+13)=X(3)
1710 POSY(P+13)=Y(3)
1720 POSX(P+14)=X(3)*(1+T1)+X(4)*T2-X(1)
1730  POSY(P+14)=Y(3)*(1+T1)+Y(4)*T2-Y(1)
1740POSX(P+15)=X(3)*T1+X(4)*T2
1750 POSY(P+15)=Y(3)*T1+Y(4)*T2
1760POSX(P+16)=X(1)
1770 POSY(P+16)=Y(1)
1780 ENDPROC
1790 DEF PROCDELAY(K)
1800 REM SHOW PIC FOR K SECS
1810 LOCAL T
1820 T=TIME
1830 REPEAT
1840 UNTIL TIME>T+100*K
1850 ENDPROC
1860 DEF FNX(C)=C*(H(1)+H(2)*C72+H(5)*C72-H(3)*C36-H(4)*C36)
1870  DEF FNY(C)=C*(H(4)*C54+H(5)*C18-H(3)*C54-H(2)*C18)
1880 DEF FND(I,J)=SQR((POSX(I)-POSX(J))^2+(POSY(I)-POSY(J))^2)
```

PLOT

Most microcomputers give graphics of very limited resolution, whereas a matrix
plotter may give pictures of much higher resolution (more distinct pixels per
line). It is thus desirable to be able to plot on the matrix plotter (straight line)
vectors from (x_1, y_1) to (x_2, y_2), which can be calculated with high accuracy.
This program does this for the Epson FX-80 printer. The program reads an
input file called VECTORS.DAT which must be written earlier and which contains the x_1, y_1, x_2, y_2 coordinates of the ends of the vectors.

```
1000 REM program name PLOT
1010 REM Plot lines from file on EPSON FX-80 printer
1020 REM input file of vectors x1,y1,x2,y2
1030 REM preceded by no. of vectors NV
1040 OPEN "I",1,"VECTORS.DAT"
1050 REM initialise printer
1060 LPRINT CHR$(27);"@";
1070 REM line width infinite
1080 WIDTH LPRINT 255
1090 REM set line height
1100 LPRINT CHR$(27);"3";CHR$(24);
1110 DEFINT I-N
1120 N=720: REM number of dots in mode 6=720
1130 DIM LROW(720)
1140 INPUT £1, NV: REM number of vectors
1150 REM data
1160 REM read in max and min of X and Y
1170 REM xmax,xmin, ymax,ymin
1180 READ XMAX,XMIN,YMAX,YMIN
1190 REM data
1200 DATA 1.340,0,1,0
1210 SX=XMAX-XMIN: SY=YMAX-YMIN
1220 LPRINT "Program PLOT"
1230 LPRINT "X-scale along paper from " XMIN; "To"; XMAX
1240 LPRINT "Y-scale across paper from "YMIN; "To"; YMAX
1250 REM read in vector list
```

```
1260 DIM X1(100),Y1(100),X2(100),Y2(100)
1270 REM put x1,x2 into ascending order
1280 FOR I=1 TO NV
1290 INPUT £1,X1(I),Y1(I),X2(I),Y2(I)
1300 IF X1(I)<=X2(I) THEN GOTO 1370
1310 TR=X1(I)
1320 X1(I)=X2(I)
1330 X2(I)=TR
1340 TR=Y1(I)
1350 Y1(I)=Y2(I)
1360 Y2(I)=TR
1370 NEXT I
1380 REM print a full-width bar
1390 LPRINT CHR$(27);"*";CHR$(6);
1400 LPRINT CHR$(N MOD 256);CHR$(INT(N/256));
1410 FOR I= 1 TO N
1420 LPRINT CHR$(7);:REM margin
1430 NEXT I
1440 REM print by lines
1450 REM 80 x 8 rows of dots
1460 FOR IL=0 TO 79
1470 REM draw margins
1480 FOR L=0 TO N: LROW(L)=0: NEXT L
1490 LROW(0)=255: LROW(1)=255: LROW (N-1)=255
1500 LROW(2)=255: LROW(N-3)=255: LROW(N-2)=255
1510 IK=1
1520 IT=0
1530 REM set up band of 8 dots
1540 FOR J= 8 TO 1 STEP -1
1550 IP=IK
1560 IT=IT+IK
1570 IK=IK*2
1580 X=SX*(8*IL+J-1)/639 + XMIN
1590 REM
1600 REM plot special functions here
1610 REM example - sine wave
1620 GOTO 1700
1630 PI=3.14159
1640 Y=.5 + .5*SIN(4*PI*X)
1650 REM convert to dots
1660 NB=INT(713*(Y-YMIN)/SY)+3
1670 IF NB<3 OR NB>716 THEN GOTO 1710
1680 LROW(NB)=LROW(NB)+IP
1690 IF LROW(NB)>IT THEN LROW(NB)=LROW(NB)-IP
1700 REM if no line goto next J
1710 REM plot straight lines
1720 FOR K=1 TO NV
1730 EPS=SX/639/1.5
1740 IF X>X2(K)+EPS THEN 2080
1750 IF X<X1(K)-EPS THEN 2080
1760 IF ABS((X1(K)-X2(K))/SX)>1/639 THEN GOTO 1870
1770 REM horizontal lines
1780 L1=INT(713*(Y1(K)-YMIN)/SY)+3
1790 L2=INT(713*(Y2(K)-YMIN)/SY)+3
1800 IF (L2-L1)=0 THEN L2=L1+1
1810 FOR NB=L1 TO L2 STEP (SGN(L2-L1))
1820 IF NB<3 OR NB>716 THEN GOTO 1850
1830 LROW(NB)=LROW(NB)+IP
1840 IF LROW(NB)>IT THEN LROW(NB)=LROW(NB)-IP
1850 NEXT NB
1860 GOTO 2080
1870 GRAD=(Y2(K)-Y1(K))/(X2(K)-X1(K))
1880 Y=GRAD*(X-X1(K)) + Y1(K)
1890 GRADP=ABS(GRAD*SX/SY)
1900 IF ABS(GRADP)=<1.51 THEN GOTO 2030
1910 REM nearer horizontal lines
1920 NC=INT(713*(Y-YMIN)/SY)+3
1930 IGRAD=INT(GRADP*360/639)
1940 LT1=NC-IGRAD
1950 LT2=NC+IGRAD
```

```
1960 IF LT2=LT1 THEN LT2=LT1+1
1970 FOR NB=LT1 TO LT2 STEP (SGN(LT2-LT1))
1980 IF NB<3 OR NB>716 THEN GOTO 2010
1990 LROW(NB)=LROW(NB)+IP
2000 IF LROW(NB)>IT THEN LROW(NB)=LROW(NB)-IP
2010 NEXT NB
2020 GOTO 2080
2030 REM nearer vertical lines
2040 NB=INT(713*(Y-YMIN)/SY)+3
2050 IF NB<3 OR NB>716 THEN GOTO 2080
2060 LROW(NB)=LROW(NB)+IP
2070 IF LROW(NB)>IT THEN LROW(NB)=LROW(NB)-IP
2080 NEXT K
2090 REM
2100 NEXT J
2110 LPRINT CHR$(27);"*";CHR$(6);
2120 LPRINT CHR$(N MOD 256); CHR$(INT(N/256));
2130 FOR K=0 TO N-1
2140 LPRINT CHR$(LROW(K));
2150 NEXT K
2160 NEXT IL
2170 REM print a full-width bar
2180 LPRINT CHR$(27);"*";CHR$(6);
2190 LPRINT CHR$(N MOD 256);CHR$(INT(N/256));
2200 FOR I=1 TO N
2210 LPRINT CHR$(224); :REM solid bar
2220 NEXT I
2230 REM re-initialise printer
2240 LPRINT CHR$(27);"@";
2250 STOP
```

```
10 REM program name STAR
20 PI=3.14159
30 N=16
40 OPEN "O",1,"VECTORS.DAT"
50 WRITE £1,N
60 FOR I=1 TO N
70 X1=.5
80 Y1=.5
90 X2=2*COS(2*PI/N*I)
100 X2=X2+X1
110 Y2=2*SIN(2*PI/N*I)
120 Y2=Y2+Y1
130 PRINT X1,Y1,X2,Y2
140 WRITE £1,X1,Y1,X2,Y2
150 NEXT I
```

```
10 REM progream name GRID
20 OPEN "O",1,"VECTORS.DAT"
30 PRINT £1,20
40 DIM X1(20),X2(20),Y1(20),Y2(20)
50 FOR I=1 TO 10
60 X1(I)=.1*I
70 X2(I)=X1(I)
80 Y1(I)=0
90 Y2(I)=1
100 Y1(I+10)=.1*I
110 Y2(I+10)=Y1(I+10)
120 X1(I+10)=0
130 X2(I+10)=1
140 NEXT I
150 FOR I=1 TO 20
160 PRINT X1(I),Y1(I),X2(I),Y2(I)
170 PRINT £1, X1(I),Y1(I),X2(I),Y2(I)
180 NEXT I
```

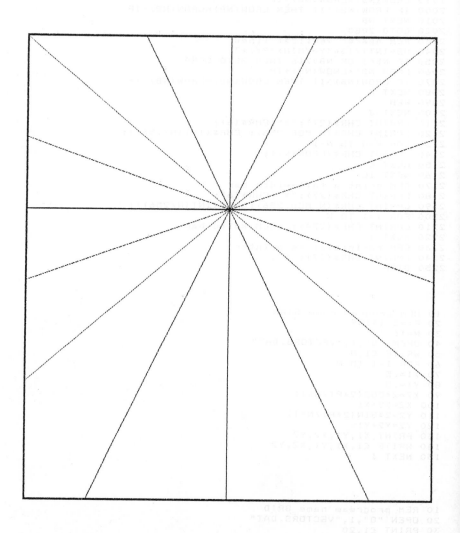

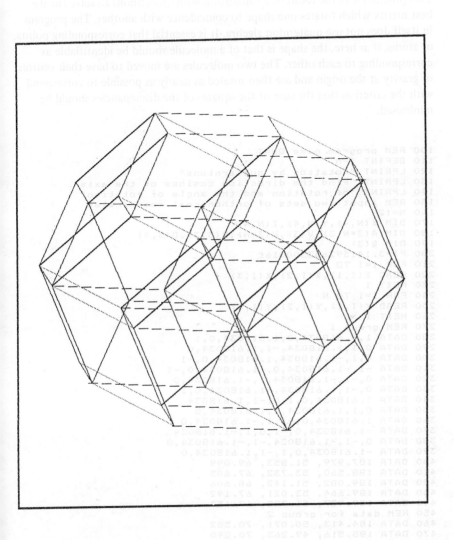

QUAT

This program uses the result of a calculation with quaternions to solve for the
best matrix which rotates one shape to coincidence with another. The program
in itself does not use quaternion algebra. It is essential that corresponding points,
or atoms, if as here, the shape is that of a molecule should be identifiable as
corresponding to each other. The two molecules are moved to have their centres
of gravity at the origin and are then rotated as nearly as possible to correspond
with the criterion that the sum of the squares of the discrepancies should be
minimised.

```
100 REM program name QUAT
110 DEFINT I-N
120 LPRINT "Rotation by quaternions"
130 LPRINT "Find the direction cosines of the axis"
140 LPRINT "of rotation and the angle of rotation"
150 REM input two sets of orthonormal coordinates
160 N=12
170 DIM X(N,4),Y(N,4),Z(N,4)
180 DIM A(3*N,3),V(3),H(3*N),C(3,3),D(3,3)
190 DIM B(3)
200 PI=3.141592653500016£
210 FOR I=1 TO N
220 READ X(I,1),X(I,2),X(I,3)
230 NEXT I
240 FOR I=1 TO N
250 READ Y(I,1),Y(I,2),Y(I,3)
260 NEXT I
270 REM group 1
280 DATA 1,1.618034,0,1.618034,0,1
290 DATA 0,1,1.618034,-1,1.618034,0
300 DATA 0,1,-1.618034,1.618034,0,-1
310 DATA -1,-1.618034,0,-1.618034,0,-1
320 DATA 0,-1,-1.618034,1,-1.618034,0
330 DATA 0,-1,1.618034,-1.618034,0,1
340 DATA 1.618034,0,1,0,-1,1.618034
350 DATA 0,1,1.618034,1,1.618034,0
360 DATA 1.618034,0,-1,1,-1.618034,0
370 DATA -1.618034,0,-1,0,1,-1.618034
380 DATA 0,-1,-1.618034,-1,-1.618034,0
390 DATA -1.618034,0,1,-1,1.618034,0
400 DATA 187.979, 51.853, 69.099
410 DATA 188.560, 53.732, 67.685
420 DATA 189.082, 51.142, 68.606
430 DATA 189.664, 53.021, 67.192
440 DATA 189.927, 51.728, 67.657
450 REM data for group 2
460 DATA 184.413, 50.071, 70.582
470 DATA 185.516, 49.263, 70.290
480 DATA 184.407, 50.873, 71.726
490 DATA 186.617, 49.252, 71.151
500 DATA 185.509, 50.864, 72.587
510 DATA 186.614, 50.053, 72.300
520 REM reduce to c.g. as origin
530 FOR I=1 TO N
540 X1=X1+X(I,1): X2=X2+X(I,2): X3=X3+X(I,3)
550 Y1=Y1+Y(I,1): Y2=Y2+Y(I,2): Y3=Y3+Y(I,3)
560 NEXT I
570 X1=X1/N: X2=X2/N: X3=X3/N
580 Y1=Y1/N: Y2=Y2/N: Y3=Y3/N
590 FOR I= 1 TO N
```

```
600 X(I,1)=X(I,1)-X1: X(I,2)=X(I,2)-X2: X(I,3)=X(I,3)-X3
610 Y(I,1)=Y(I,1)-Y1: Y(I,2)=Y(I,2)-Y2: Y(I,3)=Y(I,3)-Y3
620 NEXT I
630 REM normalise vectors to unit length
640 FOR I=1 TO N
650 S=SQR(X(I,1)*X(I,1)+X(I,2)*X(I,2)+X(I,3)*X(I,3))
660 X(I,4)=S
670 S2=SQR(Y(I,1)*Y(I,1)+Y(I,2)*Y(I,2)+Y(I,3)*Y(I,3))
680 Y(I,4)=S2
690 FOR J=1 TO 3
700 X(I,J)=X(I,J)/S: Y(I,J)=Y(I,J)/S2
710 NEXT J: NEXT I
720 FOR I=1 TO N
730 X=X(I,1): Y=X(I,2): Z=X(I,3)
740 X1=Y(I,1): Y1=Y(I,2): Z1=Y(I,3)
750 REM weight with lengths
760 W=(X(I,4)+Y(I,4))
770 A(3*I-2,1)=W*(Y1+Y)
780 A(3*I-2,2)=W*(-X1-X)
790 A(3*I-2,3)=0
800 H(3*I-2)=W*(Z1-Z)
810 A(3*I-1,1)=W*(-Z1-Z)
820 A(3*I-1,2)=0
830 A(3*I-1,3)=W*(X1+X)
840 H(3*I-1)=W*(Y1-Y)
850 A(3*I,1)=0
860 DET=DET +C(1,3)*C(2,1)*C(3,2)
870 A(3*I,2)=W*(Z1+Z)
880 DET=DET-C(1,3)*C(2,2)*C(3,1)
890 A(3*I,3)=W*(-Y1-Y)
900 H(3*I)=W*(X1-X)
910 NEXT I
920 REM form C = Atrans A
930 FOR I=1 TO 3: FOR J=1 TO 3: CX=0: FOR K=1 TO 3*N
940 CX=CX+A(K,I)*A(K,J)
950 NEXT K: C(I,J)=CX: NEXT J: NEXT I
960 REM invert explicitly. Calc. det.
970 DET=C(1,1)*C(2,2)*C(3,3)+C(1,2)*C(2,3)*C(3,1)
980 DET =DET-C(1,1)*C(2,3)*C(3,2)-C(1,2)*C(2,1)*C(3,3)
990 LPRINT "(Determinant=";DET;")"
1000 REM invert explicitly
1010 D(1,1)=(C(2,2)*C(3,3)-C(3,2)*C(2,3))/DET
1020 D(2,2)=(C(1,1)*C(3,3)-C(3,1)*C(1,3))/DET
1030 D(3,3)=(C(1,1)*C(2,2)-C(2,1)*C(1,2))/DET
1040 D(1,2)=(C(3,1)*C(2,3)-C(2,1)*C(3,3))/DET
1050 D(1,3)=(C(2,1)*C(3,2)-C(3,1)*C(2,2))/DET
1060 D(2,3)=(C(3,1)*C(1,2)-C(1,1)*C(3,2))/DET
1070 D(2,1)=D(1,2): D(3,1)=D(1,3): D(3,2)=D(2,3)
1080 FOR K=1 TO 3*N
1090 B(1)=B(1)+A(K,1)*H(K)
1100 B(2)=B(2)+A(K,2)*H(K)
1110 B(3)=B(3)+A(K,3)*H(K)
1120 NEXT K
1130 FOR J=1 TO 3
1140 V(1)=V(1)+D(1,J)*B(J)
1150 V(2)=V(2)+D(2,J)*B(J)
1160 V(3)=V(3)+D(3,J)*B(J)
1170 NEXT J
1180 REM values of l*tan(theta/2),..
1190 THETA=2*ATN(SQR(V(1)*V(1)+V(2)*V(2)+V(3)*V(3)))
1200 PI=3.1415926£
1210 LPRINT "Theta=";THETA*180/PI;" degrees"
1220 IF THETA=0 THEN STOP
1230 LPRINT "direction cosines l,m,n of axis="
1240 XL=V(1)/TAN(THETA/2)
1250 XM=V(2)/TAN(THETA/2)
1260 XN=V(3)/TAN(THETA/2)
```

```
1270 LPRINT XL;XM;XN
1280 REM rotates second mol. to coincide with the first
1290 F1=COS(THETA/2)*COS(THETA/2)-SIN(THETA/2)*SIN(THETA/2)
1300 REM check sign
1310 F2=SIN(THETA)
1320 F3=2*SIN(THETA/2)*SIN(THETA/2)
1330 FOR I=1 TO N
1340 F4=X(I,1)*XL+X(I,2)*XM+X(I,3)*XN
1350 Z(I,1)=F1*X(I,1)+F2*(XM*X(I,3)-XN*X(I,2))
1360 Z(I,2)=F1*X(I,2)+F2*(XN*X(I,1)-XL*X(I,3))
1370 Z(I,3)=F1*X(I,3)+F2*(XL*X(I,2)-XM*X(I,1))
1380 Z(I,1)=Z(I,1)+F3*F4*XL
1390 Z(I,2)=Z(I,2)+F3*F4*XM
1400 Z(I,3)=Z(I,3)+F3*F4*XN
1410 NEXT I
1420 LPRINT
1430 LPRINT "Coordinates of mol. 1 w.r.t. c.g."
1440 LPRINT "Coordinates of mol. 2 w.r.t. c.g."
1450 LPRINT "Discrepancies"
1460 FOR I=1 TO N: LPRINT
1470 LPRINT I;X(I,1)*X(I,4);X(I,2)*X(I,4);X(I,3)*X(I,4)
1480 LPRINT I;Y(I,1)*Y(I,4);Y(I,2)*Y(I,4);Y(I,3)*Y(I,4)
1490 LPRINT I;Z(I,1)*X(I,4)-Y(I,1)*Y(I,4);
1500 LPRINT Z(I,2)*X(I,4)-Y(I,2)*Y(I,4);
1510 LPRINT Z(I,3)*X(I,4)-Y(I,3)*Y(I,4)
1520 RMS=((Y(I,1)-Z(I,1))*X(I,4))^2
1530 RMS=RMS+((Y(I,2)-Z(I,2))*X(I,4))^2
1540 RMS=RMS+((Y(I,3)-Z(I,3))*X(I,4))^2
1550 NEXT I
1560 LPRINT
1570 RMS=SQR(RMS/N)
1580 LPRINT "R.M.S. discrepancy=";RMS
1590 STOP
```

```
Rotation by quaternions
Find the direction cosines of the axis
of rotation and the angle of rotation
(Determinant= 4.44878E+07 )
Theta= 72   degrees
direction cosines l,m,n of axis=
-2.11753E-09 -.525731 -.850651

Coordinates of mol. 1 w.r.t. c.g.
Coordinates of mol. 2 w.r.t. c.g.
Discrepancies

  1   1   1.61803 -9.93411E-09
  1   1.61803  0   1
  1   7.15256E-07   5.66874E-07   3.57628E-07

  2   1.61803 -2.98023E-08   1
  2  -2.98023E-08 -1   1.61803
  2   5.41233E-07  -2.38419E-07   7.15256E-07

  3  -1.98682E-08   1   1.61803
  3  -2.98023E-08   1   1.61803
  3   2.64458E-08   5.96046E-07   8.34465E-07

  4  -1   1.61803 -9.93411E-09
  4   1   1.61803 -1.98682E-08
  4  -5.96046E-08   1.07288E-06  -3.68192E-08

  5  -1.98682E-08   1  -1.61803
```

```
5    1.61803   0  -1
5    4.76837E-07   3.96812E-07  -7.15256E-07

6    1.61803  -2.98023E-08  -1
6    1  -1.61803  -1.98682E-08
6    9.53674E-07  -4.76837E-07  -3.20256E-07

7   -1  -1.61803  -9.93411E-09
7   -1.61803   0  -1
7   -7.15256E-07  -5.66874E-07  -3.57628E-07

8   -1.61803  -2.98023E-08  -1
8   -2.98023E-08   1  -1.61803
8   -4.81629E-07   2.38419E-07  -7.15256E-07

9   -1.98682E-08  -1  -1.61803
9   -2.98023E-08  -1  -1.61803
9    2.08796E-08  -5.96046E-07  -8.34465E-07

10   1  -1.61803  -9.93411E-09
10  -1  -1.61803  -1.98682E-08
10   5.96046E-08  -1.07288E-06   7.65556E-08

11  -1.98682E-08  -1   1.61803
11  -1.61803   0   1
11  -4.76837E-07  -3.68468E-07   7.15256E-07

12  -1.61803  -2.98023E-08   1
12  -1   1.61803  -1.98682E-08
12  -9.53674E-07   4.76837E-07   3.59993E-07

R.M.S. discrepancy= 3.10631E-07
```

RFT

This program, which facilitates the processing of real arrays, is again very similar to the FFT with two exceptions. Firstly, the data corresponding to the two arrays of numbers is entered as real and imaginary pairs, with the real part corresponding to the first array and the imaginary part to the second array (lines 125 to 135). Secondly, lines 355 to 365 contain the code for the recombination of the transformed data (section 10.8).

```
95  REM  RFT
96  REM  TRANSFORM FOR REAL ARRAY
100 PRINT "ENTER L WHERE N=2^L"
105 INPUT L
110 N = 2 ^ L
115 DIM X(N),Y(N),F(N),G(N)
120 PRINT "ENTER DATA AS X Y PAIRS"
125 FOR I = 1 TO N
130 INPUT X(I),Y(I)
135 NEXT I
140 M = N / 2
145 R = 2 ^ (L - 1)
150 PI = 3.141593
```

```
155 H = 1
160 A = 2 * PI / N
165 FOR P = 1 TO L
170 B = INT ((H - 1) / R)
175 GOSUB 1000
180 B = A * B
190 I = SIN (B)
195 J = COS (B)
200 FOR Q = 1 TO M
205 K = H + M
210 C = X(K) * J + Y(K) * I
215 D = Y(K) * J - X(K) * I
220 X(K) = X(H) - C
225 Y(K) = Y(H) - D
230 X(H) = X(H) + C
235 Y(H) = Y(H) + D
240 H = H + 1
245 NEXT Q
250 H = H + M
255 IF H < = N GOTO 170
260 H = 1
265 R = R / 2
270 M = M / 2
275 NEXT P
277 PRINT "TRANSFORM VALUES"
280 FOR P = 1 TO N
285 B = P - 1
290 GOSUB 1000
295 H = B + 1
320 F(P) = X(H)
325 G(P) = Y(H)
330 NEXT P
335 J = N + 2:N = N - 1
340 FOR P = 1 TO N
343 K = P + 1:M = J - K
355 R1 = 0.5 * (F(K) + F(M))
357 I1 = 0.5 * (G(K) - G(M))
360 R2 = 0.5 * (G(K) + G(M))
365 I2 = 0.5 * (F(M) - F(K))
370 PRINT R1;" ";I1
375 PRINT R2;" ";I2
380 NEXT P
385 END
1000 K = 0
1005 FOR I = 1 TO L
1010 J = INT (B / 2)
1015 K = B + INT (2 * K - 2 * J)
1020 B = J
1025 NEXT I
1030 B = K
1035 RETURN
```

```
JRUN
ENTER L WHERE N=2^L
?2
ENTER DATA AS X Y PAIRS
```

```
?1,0
?2,4
?3,5
?0,6
TRANSFORM VALUES
-2 -1.99999954
-5 2.00000012
2 0
-5 0
-2 1.99999954
-5 -2.00000012
```

SOLVE

This is a simple example of how, by solving non-linear equations by iteration, using the least squares procedure, a problem which would be very tedious to solve by exact algebraic methods can be solved numerically.

```
100 REM program name SOLVE
110 REM solution of equations
120 REM read in initial values
130 REM area S, half-perimeter P
140 REM radius of circumcircle R
150 READ S,P,R
160 REM count cycles
170 N=1
180 REM set initial (guessed) values for A,B,C
190 B=2*P/3
200 A=SQR(4*R*S/B)
210 C=SQR(A*B)
220 REM set up matrix of coefficients
230 DIM D(3,3),E(3), F(3,3), C(3)
240 GOSUB 400
250 REM solve for corrections
260 REM find inverse explicitly
270 GOSUB 770
280 FOR I=1 TO 3
290 PR=0
300 FOR J= 1 TO 3
310 PR=PR+F(I,J)*E(J)
320 NEXT J
330 PR=.2*ATN(PR*5)
340 C(I)=PR
350 NEXT I
360 A=A-C(1)
370 B=B-C(2)
380 C=C-C(3)
390 GOTO 240
400 REM subroutine to calculate
410 REM errors and coefficients
420 N=N+1
430 E(1)=S*S-P*(P-A)*(P-B)*(P-C)
440 E(2)=P -.5*(A+B+C)
450 E(3)=4*R*S-A*B*C
460 REM test errors
470 PRINT "Errors";E(1);E(2);E(3)
480 IF ABS(E(1))>.00001*P THEN 650
490 IF ABS(E(2))>.00001*P THEN 650
500 IF ABS(E(3))>.00001*P THEN 650
510 PRINT "residual errors";E(1);E(2);E(3)
520 PRINT "A=";A
```

```
530 PRINT "B=";B
540 PRINT "C=";C
550 REM check
560 P=.5*(A+B+C)
570 PRINT "Semiperimeter=";P
580 S=SQR(P*(P-A)*(P-B)*(P-C))
590 PRINT "Area=";S
600 R=.25*A*B*C/S
610 PRINT "Circumradius=";R
620 PRINT "number of cycles=";N
630 STOP
640 REM Calculate derivatives algebraically
650 D(1,1)=P*(P-B)*(P-C)
660 D(1,2)=P*(P-A)*(P-C)
670 D(1,3)=P*(P-A)*(P-B)
680 D(2,1)=-.5
690 D(2,2)=-.5
700 D(2,3)=-.5
710 D(3,1)=-B*C
720 D(3,2)=-A*C
730 D(3,3)=-A*B
740 RETURN
750 REM initial values of S,P,R
760 DATA 6,6,2.5
770 REM put explicit inverse of D(3,3) into F(3,3)
780 REM find determinant
790 DET= D(1,1)*D(2,2)*D(3,3)+D(1,2)*D(2,3)*D(3,1)
800 DET=DET+D(1,3)*D(2,1)*D(3,2)-D(1,1)*D(2,3)*D(3,2)
810 DET=DET-D(1,2)*D(2,1)*D(3,3)-D(1,3)*D(2,2)*D(3,1)
820 F(1,1)=(D(2,2)*D(3,3)-D(3,2)*D(2,3))/DET
830 F(1,2)=-(D(1,2)*D(3,3)-D(3,2)*D(1,3))/DET
840 F(2,1)=-(D(2,1)*D(3,3)-D(3,1)*D(2,3))/DET
850 F(2,2)=(D(1,1)*D(3,3)-D(1,3)*D(3,1))/DET
860 F(1,3)=(D(1,2)*D(2,3)-D(2,2)*D(1,3))/DET
870 F(3,1)=(D(2,1)*D(3,2)-D(2,2)*D(3,1))/DET
880 F(2,3)=-(D(1,1)*D(2,3)-D(2,1)*D(1,3))/DET
890 F(3,2)=-(D(1,1)*D(3,2)-D(1,2)*D(3,1))/DET
900 F(3,3)=(D(1,1)*D(2,2)-D(1,2)*D(2,1))/DET
910 RETURN
```

```
Errors-16.6825   .0955186  -.975933
Errors-15.2822   .0955257  -.508831
Errors-11.8682   .0930223   .509392
Errors-7.02974   .0792241  1.45833
Errors-1.86753   .0276241   .704014
Errors-.0392303  5.02586E-04   .011013
Errors-3.8147E-06  0   3.8147E-06
residual errors-3.8147E-06  0   3.8147E-06
A= 3
B= 4
C= 5
Semiperimeter= 6
Area= 6
Circumradius= 2.5
number of cycles= 8
```

TELLUR

This is an example of the solution of a set of simultaneous non-linear equations.
The distances of a point P from a number of datum points of known positional

coordinates are given. The calculated and observed distances of P from these points are compared and corrections to the guessed starting coordinates of P are calculated by the standard least squares methods. There may be more than the minimum number of distances necessary to fix the position of P and the resulting position is such as to minimise the sum of the squares of the discrepancies. The correction applied to P at each stage is limited. If this is not done in such problems the solution may not be smoothly approached.

```
100 REM program: name TELLUR
110 REM to find the position of a point P
120 REM given its distances to NP datum points.
130 REM number of points NP
140 NP=8
150 REM positions of datum points (x,y,z)
160 REM starting point for p
170 REM and distances
180 DIM XP(NP),YP(NP),ZP(NP),DP(NP),X(3),CX(3)
190 DIM A(3,3),B(3,3),CXX(3),DX(NP)
200 PI=3.1415926£
210 REM assume a starting position
220 X(1)=0
230 X(2)=0
240 X(3)=0
250 DIM M(NP,3)
260 REM coordinates of datum points
270 DATA 0,0,0,1,0,0,0,1,0,0,0,1
280 DATA 0,1,1,1,0,1,1,1,0,1,1,1
290 FOR I=1 TO NP
300 READ XP(I),YP(I), ZP(I)
310 NEXT I
320 FOR I= 1 TO NP
330 READ DP(I)
340 NEXT I
350 REM distances P to datum points
360 DATA 7.18309, 6.80541, 6.86733, 6.39351
370 DATA 6.03659, 5.96606, 6.47126, 5.58187
380 REM find corrections CX(3)
390 GOSUB 610
400 GOSUB 690
410 GOSUB 870
420 GOSUB 1030
430 PRINT "CORRECTIONS"
440 FOR I=1 TO 3
450 PRINT I,CX(I)
460 REM limit corrections to CMAX
470 CMAX=2
480 CX(I)=CMAX*2/PI*ATN(CX(I)*PI/2/CMAX)
490 NEXT I
500 PRINT
510 REM apply corrections
520 X(1)=X(1)+CX(1)
530 X(2)=X(2)+CX(2)
540 X(3)=X(3)+CX(3)
550 IF ABS(CX(1)) > .000001 THEN GOTO 380
560 IF ABS(CX(2))>.000001 THEN GOTO 380
570 IF ABS(CX(3)) > .000001 THEN GOTO 380
580 PRINT "POSITION OF P"
590 PRINT X(1),X(2),X(3)
600 STOP
610 REM subroutine to find differences between
620 REM observed and calculated diatances.
630 REM r.h.s. array DX(NP)
640 FOR I=1 TO NP
```

```
650 DX(I)= (XP(I)-X(1))^2+(YP(I)-X(2))^2
660 DX(I)=DX(I)+(ZP(I)-X(3))^2 -DP(I)^2
670 NEXT I
680 RETURN
690 REM Subroutine to find matrix M(NP,3)
700 REM of differentials
710 FOR I=1 TO NP
720 M(I,1)=2*(XP(I)-X(1))
730 M(I,2)=2*(YP(I)-X(2))
740 M(I,3)=2*(ZP(I)-X(3))
750 NEXT I
760 REM form array a(3,3)=Mt(3,np)*M(np,3)
770 FOR I=1 TO 3
780 FOR J=1 TO 3
790 A=0
800 FOR K=1 TO NP
810 A=A+M(K,I)*M(K,J)
820 NEXT K
830 A(I,J)=A
840 NEXT J
850 NEXT I
860 RETURN
870 REM subroutine to invert a 3 by 3 matrix explicitly
880 REM A(3,3) inverts to B(3,3)
890 DET=A(1,1)*A(2,2)*A(3,3)+A(1,2)*A(2,3)*A(3,1)
900 DET =DET+A(1,3)*A(2,1)*A(3,2)-A(1,1)*A(2,3)*A(3,2)
910 DET=DET-A(1,2)*A(2,1)*A(3,3)-A(1,3)*A(2,2)*A(3,1)
920 B(1,1)=(A(2,2)*A(3,3)-A(3,2)*A(2,3))/DET
930 B(1,2)=-(A(2,1)*A(3,3)-A(3,1)*A(2,3))/DET
940 B(1,3)=(A(2,1)*A(3,2)-A(3,1)*A(2,2))/DET
950 B(2,2)=(A(1,1)*A(3,3)-A(1,3)*A(3,1))/DET
960 B(2,3)=-(A(1,1)*A(3,2)-A(1,2)*A(3,1))/DET
970 B(3,3)=(A(1,1)*A(2,2)-A(1,2)*A(2,1))/DET
980 REM A is symmetrical
990 B(2,1)=B(4,2)
1000 B(3,1)=B(1,3)
1010 B(3,2)=B(2,3)
1020 RETURN
1030 REM calculated corrections cx(3) to x(3)
1040 FOR I = 1 TO 3
1050 CXX(I)=0
1060 FOR J=1 TO NP
1070 CXX(I)=CXX(I) + M(J,I)*DX(J)
1080 NEXT J
1090 NEXT I
1100 FOR I=1 TO 3
1110 CX(I)=0
1120 FOR J=1 TO 3
1130 CX(I)=CX(I)+B(I,J)*CXX(J)
1140 NEXT J
1150 NEXT I
1160 RETURN
```

```
CORRECTIONS
 1              -9.75764
 2              -10.1809
 3              -7.03928

CORRECTIONS
 1              -2.43664
 2              -2.8747
 3               .417847

CORRECTIONS
 1              -1.11683
 2              -1.62994
 3               3.47292
```

```
CORRECTIONS
   1            .459801
   2           -.117696
   3          5.21131

CORRECTIONS
   1         1.50461
   2          .841233
   3         5.73302

CORRECTIONS
   1         2.46136
   2         1.96516
   3         5.5412

CORRECTIONS
   1         3.22047
  .2         3.26499
   3         3.72001

CORRECTIONS
   1         2.77263
   2         3.35245
   3          .668851

CORRECTIONS
   1         1.4875
   2         2.22543
   3          -.993173

CORRECTIONS
   1          .308899
   2          .909759
   3          -.703846

CORRECTIONS
   1          -.0321722
   2          .0988312
   3          -.160441

CORRECTIONS
   1         -1.17899E-03
   2         -7.22066E-04
   3         -3.28034E-03

CORRECTIONS
   1          1.54949E-06
   2          4.34602E-08
   3         -1.02654E-06

CORRECTIONS
   1          1.28952E-07
   2         -3.48734E-07
   3         -6.6922E-08

POSITION OF P
  3.14155        2.71826        5.8599
```

TWODFT

This program computes the two-dimensional Fourier transform of a set of sampled values of FX together with the spectral components. The sampled values in this case are taken from a square array of dimension 4 (see example in chapter 10, section 10.6).

Line 100 prompts for the dimension of the square array with the values being stored in variable $F(x, y)$. For the purpose of demonstrating the program these values are computed explicitly in lines 120-145.

The real and imaginary parts of the transform are calculated as

$$F(x, y)\cos(-2\pi(ux + vy)) \text{ and } F(x, y)\sin(-2\pi(ux + vy))$$

respectively and stored in variables $A(u, v)$ and $B(u, v)$ in lines 225 and 230. These values are normalised — that is, divided by the total number of points (N^2) — and printed at line 245 for each successive value of u and v. The spectral values are determined from $\sqrt{[A(u, v)^2 + B(u, v)^2]}$ (line 300) and printed at line 310.

```
90  REM  TWODFT
95  REM  PROGRAM COMPUTES FT IN 2D
100  PRINT "INPUT ARRAY DIMENSIONS"
105  INPUT N
110 TWOPI = 6.2832 / N:M = N * N
115  DIM A(N,N),B(N,N),F(N,N)
120  FOR XX = 1 TO N STEP 3
125 X = XX - 1
130  FOR YY = 1 TO N STEP 2
135 Y = YY - 1
140 F(X,Y) = 1
145  NEXT YY: NEXT XX
180  FOR UU = 1 TO N
185 U = UU - 1
190  FOR VV = 1 TO N
195 V = VV - 1
200 A(U,V) = 0:B(U,V) = 0
205  FOR XX = 1 TO N
210 X = XX - 1
215  FOR YY = 1 TO N
220 Y = YY - 1
225 A(U,V) = F(X,Y) * COS ( - TWOPI * (U * X + V * Y)) + A(U,V)
230 B(U,V) = F(X,Y) * SIN ( - TWOPI * (U * X + V * Y)) + B(U,V)
235  NEXT YY: NEXT XX
240  PRINT "U=";U;" ";"V=";V
245  IF  ABS (A(U,V) / M) > .00001 GOTO 255
250 A(U,V) = 0
255  IF  ABS (B(U,V) / M) > .00001 GOTO 265
260 B(U,V) = 0
265  PRINT "A=";A(U,V) / M;" ";"B=";B(U,V) / M
270  NEXT VV: NEXT UU
275  PRINT "SPECTRAL VALUES"
280  FOR UU = 1 TO N
285 U = UU - 1
```

```
290  FOR VV = 1 TO N
295  V = VV - 1
300  SPEC = SQR (A(U,V) ^ 2 + B(U,V) ^ 2)
305  PRINT "U=";U;" ";"V=";V
310  PRINT SPEC
315  NEXT VV: NEXT UU
320  END
```

```
]RUN
INPUT ARRAY DIMENSIONS
?4
U=0 V=0
A=.25 B=0
U=0 V=1
A=0 B=0
U=0 V=2
A=.25 B=0
U=0 V=3
A=0 B=0
U=1 V=0
A=.125001378 B=.125
U=1 V=1
A=0 B=0
U=1 V=2
A=.125002296 B=.124999082
U=1 V=3
A=0 B=0
U=2 V=0
A=0 B=0
U=2 V=1
A=0 B=0
U=2 V=2
A=0 B=0
U=2 V=3
A=0 B=0
U=3 V=0
A=.124995867 B=-.125
U=3 V=1
A=0 B=0
U=3 V=2
A=.124994949 B=-.125000918
U=3 V=3
A=0 B=0
SPECTRAL VALUES
U=0 V=0
4
U=0 V=1
0
U=0 V=2
4
U=0 V=3
0
U=1 V=0
2.82844271
U=1 V=1
0
U=1 V=2
2.82844271
U=1 V=3
0
U=2 V=0
0
```

```
U=2 V=1
0
U=2 V=2
0
U=2 V=3
0
U=3 V=0
2.82838037
U=3 V=1
0
U=3 V=2
2.82838037
U=3 V=3
0
```

VORON

This is a program for the BBC computer (with a colour monitor) to investigate
the geometrical distribution of domains or 'zones of influence' about eight
points whose coordinates are specified. Any given point on the screen is allocated to the point to which it is nearest (and is given an appropriate colour).
Test points are simply chosen at random and the picture builds up. To get a
complete picture (at a greater cost in waiting time) all pixels on the screen can
be scanned in turn. A function other than the distance may be used. Using the
distance gives 'frontiers' which are straight lines and this may not be the case
for other criteria. The program illustrates how an approximate picture may be
quickly obtained by sampling at random.

```
 10 REM STATISTICAL VORONOI
 20 MODE 2
 30 CLS
 40 READ N%
 50 REM NUMBER OF POINTS
 60 DATA 8
 70 REM READ IN X,Y AND RADIUS
 80 DIM X%(N%),Y%(N%),R%(N%)
 90 FOR I%=1 TO N%
100 READ X%(I%),Y%(I%),R%(I%)
110 GCOL 0,9
120 PLOT 69,X%(I%),Y%(I%)
130 NEXT I%
140 DATA 100,100,200
150 DATA 300,300,200
160 DATA 500,500,200
170 DATA 700,700,200
180 DATA 900,900,200
190 DATA 1279,0,10
200 DATA 200,800,10
210 DATA 640,512,2
220 REM MAIN CYCLE TO CHOOSE RANDOM POINTS
230 FOR I%=1 TO 10000
240 X%=RND(1279)
250 Y%=RND(1023)
260 T=10000000
270 P%=0
280 FOR J%=1 TO N%
```

```
290 T1=FNDIST(X%(J%),Y%(J%),R%(J%))
300 IF T1<T THEN P%=J%: T=T1
310 NEXT
320 GCOL 0, (P%-1) MOD 7 + 1
330 PLOT 69,X%,Y%
340 NEXT
350 STOP
360 END
370 DEF FNDIST(A%,B%,R%)=((A%-X%)^2+(B%-Y%)^2-R%*R%)
```

WALSH

This program calculates the WALSH transform of a given function f(x) for
N = 8.

The values of $b_k(z)$ in the transformation kernel are stored in the array B(I, J) where I corresponds to k and J corresponds to z.

The table of values for g(x, u) is calculated in the loop (lines 155–165).

Each value for f(x) is requested at line 185 and the final values for the transform are calculated in the loop (lines 210–220) with the results being printed at line 230.

```
95  REM  WALSH
100  DIM F(8),G(8,8),B(8,8),SW(8)
105  FOR I = 0 TO 7
110  FOR J = 0 TO 7
115  B(I,J) = 0
120  NEXT J: NEXT I
125  B(0,1) = 1:B(1,2) = 1:B(0,3) = 1:B(1,3) = 1
130  B(2,4) = 1:B(0,5) = 1:B(2,5) = 1:B(1,6) = 1
135  B(2,6) = 1:B(0,7) = 1:B(1,7) = 1:B(2,7) = 1
140  FOR X = 0 TO 7
145  FOR U = 0 TO 7
150  G(X,U) = 1
155  FOR I = 0 TO 2
157  M = 2 - I
160  G(X,U) = (( - 1) ^ (B(I,X) * B(M,U))) * G(X,U)
165  NEXT I
175  NEXT U: NEXT X
180  FOR X = 0 TO 7
185  PRINT "ENTER FX"
190  INPUT F(X)
195  NEXT X
200  FOR U = 0 TO 7
205  SW(U) = 0
210  FOR X = 0 TO 7
215  SW(U) = SW(U) + F(X) * G(X,U)
220  NEXT X
225  SW(U) = SW(U) / 8
230  PRINT U;" ";SW(U)
235  NEXT U
240  END


JRUN
ENTER FX
?1
ENTER FX
?2
ENTER FX
?3
ENTER FX
?4
ENTER FX
?0
ENTER FX
?0
ENTER FX
?0
ENTER FX
```

```
?0
0 1.25
1 1.25
2 -.5
3 -.5
4 -.25
5 -.25
6 0
7 0
```

APL programs

We give here the three programs, DETER, GENINV and JACOBI in the form
of APL used on microcomputers where the special symbols are replaced by
ordinary words. These complement the ordinary system functions and bring
up the version to the facilities provided by APL2. We also give, as an example
of a much more complicated calculation, the program LANGLET which calculates the radius and centre of a sphere which is to be inscribed between
four spheres of arbitrary position and radius.

DETER

```
        defn det<,>
        defn DET A;IJ;MX;BX;PX;IX;JX;UX;CX
<1>     DX is 1
<2>     LAB1:IJ is 2 size size A
<3>     NX is IJ<1>
<4>     goto LAB3 if IJ<1>ne IJ<2>
<5>     goto LAB2 if NX eq 1
<6>     MX is max on(max on abs A)
<7>     BX is(abs A)in MX
<8>     PX is(,BX)index 1
<9>     IX is max PX/NX
<10>    JX is PX - (IX - 1)*NX
<11>    DX is DX*A<IX;JX>
<12>    CX is(A<IX;>outer.*A<;JX>)/A<IX,JX>
<13>    CX is trans CX
<14>    A is A - CX
<15>    UX is IJ size 1
<16>    UX<IX;>is 0
<17>    UX<;JX>is 0
<18>    A is(IJ - 1 1)size((,UX)on,A)
<19>    goto LAB1
<20>    LAB2:DX is DX*A
<21>    prompt is'Determinant
<22>    DX
<23>    LAB3:
        defn
<24>    defn
```

```
        det a
Determinant
        0
        a
  9.65685E0        0.00000E0        7.06683E-16    -5.41196E-1
5.41196E-1
```

```
          5.22625E0      2.23044E0
     0.00000E0      0.00000E0      0.00000E0      0.00000E0
0.00000E0
               0.00000E0      0.00000E0
     1.29237E-15   0.00000E0     -1.65685E0     -1.30656E0
1.30656E0
               2.16478E0      1.58513E-1
    -5.41196E-1    0.00000E0     -1.30656E0     -4.00000E0      -
2.00000E0
               1.41421E0      0.00000E0
     5.41196E-1    0.00000E0      1.30656E0     -2.00000E0      -
4.00000E0
              -1.41421E0      0.00000E0
     5.22625E0     0.00000E0      2.16478E0      1.41421E0      -
1.41421E0
               0.00000E0      1.00000E0
     2.23044E0     0.00000E0      1.58513E-1     0.00000E0
0.00000E0
               1.00000E0      0.00000E0
```

GENINV

```
         DEFN GENINV <,>
         defn GENINV A;SX;TX;UX;IX;EX;EX2;SKX;PX;SK1X;TRX
<1>      SX is 2 size size A
<2>      TX is SX<1>size 0
<3>      UX is 2 size SX<1>
<4>      IX is UX size 1,TX
<5>      EX is 0.01/(+on(+on abs A))
<6>      EX2 is 1E-13
<7>      SKX is EX*trans A
<8>   LAB1:PX is A+.*SKX
<9>      SK1X is SKX+.*((2*IX) - PX)
<10>     TRX is+on(+on IX*PX)
<11>     TRX
<12>     goto LAB2 if abs(1 abs TRX)lt EX2
<13>     SKX is SK1X
<14>     goto LAB1
<15>  LAB2:ANS is SK1X
<16>     !puts geninv of a into ans
<17>     ANS
         defn
<18>  defn
```

```
         geninv a
0.0342105
0.0676841
0.132494
0.254052
0.468473
0.805962
1.24499
1.69671
2.11584
2.50881
2.89497
3.301
3.64636
3.88404
3.9866
3.99982
4
4
```

0.56	−0.44	−0.02	−0.02
−0.44	0.56	−0.02	−0.02
−0.02	−0.02	0.34	−0.16
−0.02	−0.02	−0.16	0.34

```
       a+.*ans
 1.00000E0      −2.42861E-16    2.08167E-17    2.08167E-17
−2.42861E-16     1.00000E0      2.08167E-17    2.08167E-17
−6.24500E-17    −6.24500E-17    1.00000E0      4.16334E-17
−6.24500E-17    −6.24500E-17    4.16334E-17    1.00000E0
```

JACOBI

```
      defn jacobi <,>
      defn JACOBI A;NX;CX;BX;MX;QX;PX;IX;JX;RX;TX;DX
<1>   !finds eigenvalues and vectors of a symmetrical matri
x
<2>   NX is 2 size size A
<3>   goto LAB1 if NX<1>ne NX<2>
<4>   CX is NX size 1,NX<1>on 0
<5>   CX is A ne trans A
<6>   'NOTSYMMETRICAL'if(+on+on CX)ne 0
<7>   'ORIGINAL MATRIX'
<8>   A
<9>   CX is NX size 1,NX<1>on 0
<10>  DX is CX
<11>  !find largest off-diag. element
<12>  'Largest off-diagonal element'
<13>  LAB5:BX is A − CX*A
<14>  MX is max on(max on abs BX)
<15>  goto LAB3 if MX lt 1E-12
<16>  QX is(abs BX)in MX
<17>  PX is(,QX)index 1
<18>  IX is max PX/NX<1>
<19>  JX is PX −.(IX − 1)*NX<1>
<20>  A<IX;JX>
<21>  RX is NX size 1,NX<1>on 0
<22>  goto LAB2 if(abs(A<IX;IX> − A<JX;JX>))gt 1E-12
<23>  TX is trig 0.25
<24>  goto LAB4
<25>  LAB2:TX is 2*A<IX;JX>/(A<IX;IX> − A<JX;JX>)
<26>  TX is 0.5*(-3 trig TX)
<27>  LAB4:TX is − TX
<28>  RX<IX;IX>is 2 trig TX
<29>  RX<JX;JX>is 2 trig TX
<30>  RX<JX;IX>is 1 trig TX
<31>  RX<IX;JX>is − 1 trig TX
<32>  A is RX+.*A+.*trans RX
<33>  DX is RX+.*DX
<34>  goto LAB5
<35>  LAB3:
<36>  'diagonalised  matrix'
<37>  A
<38>  'eigenvectors are roxw'
<39>  'EIGVEC+.*A+.*trans EIGVEC diagonalises'
<40>  'original matrix A'
<41>  EIGVEC is DX & 'EIGVEC'
<42>  DX
<43>  ' '
```

```
<44>    'Eigenvalues'
<45>    EIGVAL is(NX<1>,1)size 1 1 trans A
<46>    EIGVAL
<47>    LAB1:
        defn
<48>    defn
```

```
ORIGINAL MATRIX
  0  1  3  4  3  1  1
  1  0  1  3  4  3  1
  3  1  0  1  3  4  1
  4  3  1  0  1  3  1
  3  4  3  1  0  1  1
  1  3  4  3  1  0  1
  1  1  1  1  1  1  0
Largest off-diagonal element
4
4
4
4
5.22625
2.44426
-2
-2.13908
2.71061
0.713445
-0.521831
-0.0751447
0.030957
0.0299222
0.00541456
-0.000852847
0.000571476
-0.000568218
-0.000560469
-2.69475E-5
3.99836E-6
6.62836E-7
-3.03886E-7
-3.82972E-8
3.18511E-8
2.85063E-11
-1.95218E-11
1.59081E-12
diagonalised    matrix
  1.24807E1        0.00000E0       -4.83284E-16      0.00000E0      -
1.07219E-21
          1.69165E-15    4.15138E-28
  0.00000E0        0.00000E0        0.00000E0        0.00000E0
0.00000E0
          0.00000E0        0.00000E0
  7.09975E-30      0.00000E0       -2.35451E-16      0.00000E0
3.74538E-16
          3.57814E-16    8.68103E-24
  0.00000E0        0.00000E0        0.00000E0       -6.00000E0
0.00000E0
          0.00000E0        0.00000E0
 -8.18556E-17      0.00000E0        2.46988E-16      0.00000E0
3.44042E-18
         -8.89246E-26    6.40332E-28
  1.47594E-15      0.00000E0        1.41817E-22      0.00000E0      -
7.94477E-17
         -6.00000E0       -1.88866E-15
```

```
   4.18982E-17    0.00000E0      9.55342E-17    0.00000E0     -
1.56679E-17
          -1.93559E-15   -4.80741E-1
eigenvectors are roxw
EIGVEC+.*A+.*trans EIGVEC diagonalises
original matrix A
EIGVEC
   4.00606E-1     4.00606E-1     4.00606E-1     4.00606E-1
4.00606E-1
          4.00606E-1     1.92588E-1
  -5.00000E-1     5.00000E-1     0.00000E0     -5.00000E-1
5.00000E-1
          0.00000E0      0.00000E0
  -7.58278E-2    -4.40676E-1     6.98928E-1    -4.40676E-1    -
7.58278E-2
          3.34080E-1    -3.92036E-16
  -5.00000E-1    -5.00000E-1     0.00000E0      5.00000E-1
5.00000E-1
          0.00000E0      0.00000E0
   4.94217E-1    -2.36230E-1     1.07237E-1    -2.36230E-1
4.94217E-1
         -6.23210E-1     1.57499E-16
   2.88675E-1    -2.88675E-1    -5.77350E-1    -2.88675E-1
2.88675E-1
          5.77350E-1    -3.30481E-16
  -7.86235E-2    -7.86235E-2    -7.86235E-2    -7.86235E-2    -
7.86235E-2
         -7.86235E-2     9.81280E-1

Eigenvalues
   1.24807E1
   0.00000E0
  -2.35451E-16
  -6.00000E0
   3.44042E-18
  -6.00000E0
  -4.80741E-1
```

LANGLET

```
      defn LANGLET
<1>   !To find the radius and centre of a sphere
<2>   ! inscribed between four others with given
<3>   ! radii and coords. of centres.
<4>   ! T is 4 4 Matrix of 4 rows of r, x, y, z
<5>   G is( - Q<;1>),0 1 drop Q is T*T
<6>   S is 0.5*+on -1 0 drop G - 1 rotate<1>G
<7>   P is -1 drop(1 rotate T<;1>) - T<;1>
<8>   M is matdiv M is -1 0 drop U - 1 rotate<1>U is 0 1 d
rop T
<9>   S is M+.*S
<10>  P is M+.*P
<11>  V1 is 1 drop T<1;>
<12>  R1 is T<1;1>
<13>  D is S - V1
<14>  A is+on -1,P*P
<15>  BP is(D+.*P) - R1
<16>  C is(D,R1)+.*D. - R1
<17>  DEL is(BP*BP) - A*C
<18>  ! THEN, IF DEL >=0
<19>  R is(( - BP) - DEL exp 0.5)/A
<20>  V is S+P*R
<21>  'COORDINATES OF CENTRE' & V
```

```
<22>   'RADIUS' & R
<23>   ! G.A.LANGLET, ACTA CRYST., A35, 836-837, 1979.
       defn
<24>   defn


       T
     0.5              1              1              0
     0.5              1              0              1
     0.5              0              1              1
     0.5              0              0              0


       LANGLET
COORDINATES OF CENTRE
0.5 0.5 0.5
RADIUS
0.366025
```

References

E. O. Brigham (1974). *The Fast Fourier Transform*, Prentice-Hall, Englewood Cliffs, New Jersey.

W. Brostow, J-P. Dussault and B. L. Fox (1978). *J. Comp. Phys.*, **29**, 81–92.

J. W. Cooley, P. A. W. Lewis and P. D. Welch (1969). 'The Fast Fourier Transform and its Applications', *IEEE Trans. Educ.*, **E-12**(1), 27–34.

D. deSolla Price (1974). *Trans. Amer. Phil. Soc.*, **62**(2), 1–70.

J. Felsenstein, S. Sawyer and R. Kochim (1982). 'An efficient method for matching nucleic acid sequences', *Nucleic Acids Research*, **10**(1).

J. L. Finney (1979). *J. Comp. Phys.*, **32**, 137–143.

B. Hague (1970). *An Introduction to Vector Analysis*, Methuen, London.

M. Hasegawa and M. Tanemura (1980). *J. Theor. Biol.*, **82**, 477–496

B. Julesz (1965). *Scientific American*, February, pp. 38–48.

G. A. Korn and T. M. Korn (1968). *Mathematical Handbook for Scientists and Engineers*, 2nd edn, McGraw-Hill, New York.

A. L. Mackay (1984). *Acta Cryst.*, **A40**, 165–166.

J. C. Maxwell (1869). *Scientific Papers of J. C. Maxwell*, pp. 514–525 (Chelsea Reprint Co., New York).

G. F. Oster and C. A. Desoer (1971). *J. Theor. Biol.*, **32**, 219–241.

Poole and Borchers (1979). *Some Common Basic Programs*, Osborne.

J. S. Rollett (ed.) (1965). *Computing Methods in Crystallography*, Pergamon, Oxford.

J. Rooney (1977). *Environ. Planning*, **B4**, 185–210.

H. A. Simon (1969). *The Science of the Artificial*, MIT Press, Cambridge, Massachusetts.

C. E. Weatherburn (1924). *Advanced Vector Analysis*, Bell, London.

C. E. Weatherburn (1935). *Elementary Vector Analysis*, Bell, London.

C. E. Weatherburn (1939). *Differential Geometry of Three Dimensions*, Cambridge University Press.

C. Zwicker (1950). *Advanced Plane Geometry*, North-Holland, Amsterdam.

Further reading

I. O. Angell and G. Griffith, *High-resolution Computer Graphics using FORTRAN 77*, Macmillan, London, 1987.

I. O. Angell and B. J. Jones, *Advanced Graphics with the BBC Model B Microcomputer*, Macmillan, London, 1983.

I.O. Angell and B. J. Jones, *Advanced Graphics with the Sinclair ZX Spectrum*, Macmillan, London, 1983.

A. Bowyer, *Comp. J.*, **24** (1981) 162–166. [Voronoi dissections].

A. Bowyer and J. Woodwark, *A Programmer's Geometry*, Butterworths, London, 1983.

R. N. Bracewell, *The Fourier Transform and its Applications*, McGraw-Hill Kogakusha.

I. N. Bronshtein and K. A. Semendyaev, *Handbook of Mathematics* [in various languages], Moscow, 1967 etc.

E. B. Brown, 'The Surface Proteins of Influenza Virus', Thesis for Diploma of the Royal Microscopical Society, Oxford.

H. S. M. Coxeter, *Introduction to Geometry*, Wiley, New York, 1961.

H. M. Cundy and A. P. Rollett, *Mathematical Models*, Clarendon Press, Oxford, 1951.

H. B. Dwight, *Tables of Integrals and Other Mathematical Data*, Macmillan Inc., New York, 1949.

Euclid [of Alexandria] (*ca.* 300 BC), *Thirteen Books of the Elements* (Dover Reprint, New York).

W. Fischer and E. Koch, *N. Jb. Miner. Mh.*, (1973) 252–273 and 361–380. [Voronoi dissections].

W. Fisher and E. Koch, *Zeit. f. Krist.*, (1979) 245–260. [Voronoi dissections].

J. D. Foley and A. van Dam, *Principles of Interactive Computer Graphics*, Addison-Wesley, Reading, Massachusetts, 1982.

F. C. Frank and J. S. Kaspar, *Acta Cryst.*, **11** (1958) 184–190. [Voronoi dissections].

R. C. Gonzalez and P. Wintz, *Digital Image Processing*, Addison-Wesley, Advanced Book Program/World Science Division, Reading, Massachusetts, 1977.

M. J. Geisow and A. N. Barrett (eds), *Computing in Biological Science*, Elsevier, Amsterdam, 1983.

L. Gilman and A. J. Rose, *APL – An Interactive Approach*, 3rd edn, Wiley, New York.

P. J. Green and R. Sibson, *Comp. J.*, **21** (1978) 168–173. [Voronoi dissections].

A. L. Loeb, *Space Structures*, Addison-Wesley, Reading, Massachusetts, 1976.

A. L. Mackay, *J. Microscopy*, **95** (1972) 217–227. [Voronoi dissections].

W. M. Newman and R. F. Sproull, *Principles of Interactive Computer Graphics*, 2nd edn, McGraw-Hill, London, 1979.

R. P. Paul, *Robot Manipulators*, MIT Press, Cambridge, Massachusetts, 1981.

QL/APL Reference Manual, MicroAPL Ltd (Unit 1F, Nine Elms Industrial Estate, 87 Kirtling Street, London SW8 5BP), 1985.

G. Stephenson, *Mathematical Methods for Science Students*, Longman, London, 1961.

D. F. Watson, *Comp. J.*, **24** (1981) 167–172. [Voronoi dissections].

R. Williams, *The Geometrical Foundation of Natural Structure*, Dover Publications, New York, 1972.

Index